SKILLS MASTERY

This Book Includes:

- Practice questions to help students master topics assessed on the the PARCC and Smarter Balanced Tests
  - ▸ Operations & Algebraic Thinking
  - ▸ Number & Operations in Base Ten
  - ▸ Number & Operations – Fractions
  - ▸ Measurement & Data
  - ▸ Geometry
- Detailed answer explanations for every question
- Strategies for building speed and accuracy
- Content aligned with the Common Core State Standards

Plus access to Online Workbooks which include:

- Hundreds of practice questions
- Self-paced learning and personalized score reports
- Instant feedback after completion of the workbook

Complement Classroom Learning All Year

Using the Lumos Study Program, parents and teachers can reinforce the classroom learning experience for children. It creates a collaborative learning platform for students, teachers and parents.

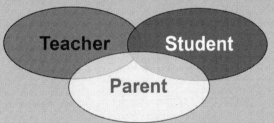

**Used in Schools and Libraries
To Improve Student Achievement**

Lumos Learning

Common Core Practice - Grade 4 Math: Workbooks to Prepare for the PARCC or Smarter Balanced Test

Contributing Author	-	Jessica Fisher
Contributing Author	-	Wilma Muhammad
Curriculum Director	-	Marisa Adams
Executive Producer	-	Mukunda Krishnaswamy
Designer	-	Mirona Jova
Database Administrator	-	R. Raghavendra Rao

COPYRIGHT ©2015 by Lumos Information Services, LLC. **ALL RIGHTS RESERVED.** No part of this work covered by the copyright hereon may be reproduced or used in any form or by an means graphic, electronic, or mechanical, including photocopying, recording, taping, Web distribution or information storage and retrieval systems- without the written permission of the publisher.

ISBN-10: 1940484448

ISBN-13: 978-1-940484-44-0

Printed in the United States of America

For permissions and additional information contact us

Lumos Information Services, LLC
PO Box 1575, Piscataway, NJ 08855-1575
http://www.LumosLearning.com

Email: support@lumoslearning.com
Tel: (732) 384-0146
Fax: (866) 283-6471

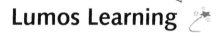

Table of Contents

Introduction 1
How to use this book effectively 2
How to access the Lumos Online workbooks 4
Lumos StepUp™ Mobile App FAQ 5
Lumos SchoolUp™ Mobile App FAQ 6

Operations & Algebraic Thinking 7
Number Sentences (4.OA.A.1) 7
Real World Problems (4.OA.A.2) 10
Multi-Step Problems (4.OA.A.3) 12
Number Theory (4.OA.A.4) 14
Answer Key and Detailed Explanations 17

Number & Operations in Base Ten 23
Place Value (4.NBT.A.1) 23
Compare Numbers and Expanded Notation (4.NBT.A.2) 26
Rounding Numbers (4.NBT.A.3) 28
Addition & Subtraction (4.NBT.B.4) 30
Multiplication (4.NBT.B.5) 33
Division (4.NBT.B.6) 36
Answer Key and Detailed Explanations 39

Numbers & Operations – Fractions 47
Equivalent Fractions (4.NF.A.1) 47
Compare Fractions (4.NF.A.2) 51
Adding and Subtracting Fractions (4.NF.B.3.A) 55
Adding and Subtracting Fractions through Decompositions (4.NF.B.3.B) 57
Adding and Subtracting Mixed Numbers (4.NF.B.3.C) 58
Adding and Subtracting Fractions in Word Problems (4.NF.B.3.D) 60
Multiplying Fractions (4.NF.B.4.A) 62
Multiplying Fractions by a Whole Number (4.NF.B.4.B) 64
Multiplying Fractions in Word Problems (4.NF.B.4.C) 65

10 to 100 Equivalent Fractions (4.NF.C.5) 67
Convert Fractions to Decimals (4.NF.C.6) 69
Compare Decimals (4.NF.C.7) 72
Answer Key and Detailed Explanations 75

Measurement & Data 86
Units of Measurement (4.MD.A.1) 86
Measurement Problems (4.MD.A.2) 89
Perimeter & Area (4.MD.B.3) 92
Representing and Interpreting Data (4.MD.B.4) 95
Angle Measurement (4.MD.C.5.A) 110
Measuring Turned Angles (4.MD.C.5.B) 115
Measuring and Sketching Angles (4.MD.C.6) 117
Adding and Subtracting Angle Measurements (4.MD.C.7) 119
Answer Key and Detailed Explanations 121

Geometry 131
Points, Lines, Rays, and Segments (4.G.A.1) 131
Angles (4.G.A.1) 134
Classifying Plane (2-D) Shapes (4.G.A.2) 137
Symmetry (4.G.A.3) 140
Answer Key and Detailed Explanations 143

Introduction

The Common Core State Standards Initiative (CCSS) was created from the need to have more robust and rigorous guidelines, which could be standardized from state to state. These guidelines create a learning environment where students will be able to graduate high school with all skills necessary to be active and successful members of society, whether they take a role in the workforce or in some sort of post-secondary education.

Once the CCSS were fully developed and implemented, it became necessary to devise a way to ensure they were assessed appropriately. To this end, states adopting the CCSS have joined one of two consortia, either PARCC or Smarter Balanced.

Why Practice by Standard?

Each standard, and substandard, in the CCSS has its own specific content. Taking the time to study and practice each one individually can help students more adequately understand the CCSS for their particular grade level. Additionally, students have individual strengths and weaknesses. Being able to practice content by standard allows them the ability to more deeply understand each standard and be able to work to strengthen academic weaknesses.

How Can the Lumos Study Program Prepare Students for Standardized Tests?

Since the fall of 2014, student mastery of Common Core State Standards are being assessed using standardized testing methods. At Lumos Learning, we believe that yearlong learning and adequate practice before the actual test are the keys to success on these standardized tests. We have designed the Lumos study program to help students get plenty of realistic practice before the test and to promote yearlong collaborative learning.

This is a Lumos tedBook™. It connects you to Online Workbooks and additional resources using a number of devices including android phones, iPhones, tablets and personal computers. Each Online Workbook will have some of the same questions seen in this printed book, along with additional questions. The Lumos StepUp® Online Workbooks are designed to promote yearlong learning. It is a simple program students can securely access using a computer or device with internet access. It consists of hundreds of grade appropriate questions, aligned to the new Common Core State Standards. Students will get instant feedback and can review their answers anytime. Each student's answers and progress can be reviewed by parents and educators to reinforce the learning experience.

How to use this book effectively

The Lumos Program is a flexible learning tool. It can be adapted to suit a student's skill level and the time available to practice before standardized tests. Here are some tips to help you use this book and the online resources effectively:

Students

- The standards in each book can be practiced in the order designed, or in the order of your own choosing.
- Complete all problems in each workbook.
- Use the Online workbooks to further practice your areas of difficulty and complement classroom learning.
- Download the Lumos StepUp® app using the instructions provided to have anywhere access to online resources.
- Practice full length tests as you get closer to the test date.
- Complete the test in a quiet place, following the test guidelines. Practice tests provide you an opportunity to improve your test taking skills and to review topics included in the CCSS related standardized test.

Parents

- Familiarize yourself with your state's consortium and testing expectations.
- Get useful information about your school by downloading the Lumos SchoolUp™ app. Please follow directions provided in "How to download Lumos SchoolUp™ App" section of this chapter.
- Help your child use Lumos StepUp® Online Workbooks by following the instructions in "How to access the Lumos Online Workbooks" section of this chapter.
- Help your student download the Lumos StepUp® app using the instructions provided in "How to download the Lumos StepUp® Mobile App" section of this chapter.
- Review your child's performance in the "Lumos Online Workbooks" periodically. You can do this by simply asking your child to log into the system online and selecting the subject area you wish to review.
- Review your child's work in each workbook.

Teachers

- You can use the Lumos online programs along with this book to complement and extend your classroom instruction.

- Get a Free Teacher account by using the respective states specific links and QR codes below:

This Lumos StepUp® Basic account will help you:

- Create up to 30 student accounts.
- Review the online work of your students.
- Easily access CCSS.
- Create and share information about your classroom or school events.

NOTE: There is a limit of one grade and subject per teacher for the free account.

- Download the Lumos SchoolUp™ mobile app using the instructions provided in "How can I Download the App?" section of this chapter.

How to Access the Lumos Online Workbooks

First Time Access:

In the next screen, click on the "New User" button to register your user name and password.

Subsequent Access:

After you establish your user id and password for subsequent access, simply login with your account information.

What if I buy more than one Lumos Study Program?

Please note that you can use all Online Workbooks with one User ID and Password. If you buy more than one book, you will access them with the same account.

Go back to the **http://www.lumoslearning.com/book** link and enter the access code provided in the second book. In the next screen simply login using your previously created account.

Lumos StepUp® Mobile App FAQ For Students

What is the Lumos StepUp® App?
It is a FREE application you can download onto your Android smart phones, tablets, iPhones, and iPads.

What are the Benefits of the StepUp® App?
This mobile application gives convenient access to Practice Tests, Common Core State Standards, Online Workbooks, and learning resources through your smart phone and tablet computers.
- Eleven Technology enhanced question types in both MATH and ELA
- Sample questions for Arithmetic drills
- Standard specific sample questions
- Instant access to the Common Core State Standards
- Jokes and cartoons to make learning fun!

Do I Need the StepUp® App to Access Online Workbooks?
No, you can access Lumos StepUp® Online Workbooks through a personal computer. The StepUp® app simply enhances your learning experience and allows you to conveniently access StepUp® Online Workbooks and additional resources through your smart phone or tablet.

How can I Download the App?
Visit **lumoslearning.com/a/stepup-app** using your smart phone or tablet and follow the instructions to download the app.

QR Code
for Smart Phone
Or Tablet Users

Lumos SchoolUp™ Mobile App FAQ For Parents and Teachers

What is the Lumos SchoolUp™ App?

It is a FREE App that helps parents and teachers get a wide range of useful information about their school. It can be downloaded onto smartphones and tablets from popular App Stores.

What are the Benefits of the Lumos SchoolUp™ App?

It provides convenient access to
- School "Stickies". A Sticky could be information about an upcoming test, homework, extra curricular activities and other school events. Parents and educators can easily create their own sticky and share with the school community.
- Common Core State Standards.
- Educational blogs.
- StepUp™ student activity reports.

How can I Download the App?

Visit **lumoslearning.com/a/schoolup-app** using your smartphone or tablet and follow the instructions provided to download the App. Alternatively, scan the QR Code provided below using your smartphone or tablet computer.

QR Code
for Smart Phone
Or Tablet Users

Name _____   Date _____

Operations & Algebraic Thinking

Section 1: Represent and solve problems involving multiplication and division.

Number Sentences (4.OA.A.1)

1. Andrew is twice as old as his brother, Josh. Which equation could be used to figure out Andrew's age if Josh's age, j, is unknown?

   Ⓐ a = j + 2
   Ⓑ a = j ÷ 2
   Ⓒ j = a + 2
   Ⓓ a = 2 × j

2. Mandy bought 28 marbles. She wants to give the same number of marbles to each of her four friends. What equation or number sentence would she use to find the number of marbles each friend will get?

   Ⓐ 28 - 4 = n
   Ⓑ 28 ÷ 4 = n
   Ⓒ 28 + 4 = n
   Ⓓ 28 - 4 = n

3. What number does n represent?
   3 + 6 + n = 22

   Ⓐ n = 9
   Ⓑ n = 13
   Ⓒ n = 18
   Ⓓ n = 31

Name _____   Date _____

4. Cindy's mother baked cookies for the school bake sale. Monday she baked 4 dozen cookies. Tuesday she baked 3 dozen cookies. Wednesday she baked 4 dozen cookies. After she finished baking Thursday afternoon, she took 15 dozen cookies to the bake sale. Which equation shows how to determine the amount of cookies that she baked on Thursday?

   Ⓐ $4 + 3 + 4 + n = 15$
   Ⓑ $4 + 3 + 4 = n$
   Ⓒ $4 \times 3 \times 4 \times n = 15$
   Ⓓ $15 \div 11 = n$

5. There are 9 students in Mrs. Whitten's class. She gave each student the same number of popsicle sticks. There were 47 popsicle sticks in her bag. To decide how many sticks each student received, Larry wrote the following number sentence: $47 \div 9 = n$. How many popsicle sticks were left in the bag after dividing them evenly among the 9 students?

   Ⓐ 0
   Ⓑ 2
   Ⓒ 3
   Ⓓ 4

6. Sixty-three students visited the science exhibit. The remainder of the visitors were adults. One hundred forty-seven people visited the science exhibit in all.
   How would you determine how many of the visitors were adults?

   Ⓐ $63 + 147 = n$
   Ⓑ $147 \div 63 = n$
   Ⓒ $147 \div n = 63$
   Ⓓ $63 + n = 147$

7. Donald bought a rope that was 89 feet long. To divide his rope into 11 foot long sections, he solved the following problem: $89 \div 11 = n$. How many feet of rope were left over?

   Ⓐ 0 feet
   Ⓑ 1 foot
   Ⓒ 2 feet
   Ⓓ 3 feet

8. If 976 − n = 325 is true, which of the following equations is NOT true?

   Ⓐ 976 + 325 = n
   Ⓑ 976 − 325 = n
   Ⓒ n + 325 = 976
   Ⓓ 325 + n = 976

9. Mary has $54. Jack has n times as much money as Mary does. The total amount of money Jack has is $486. What is n?

   Ⓐ 19
   Ⓑ 29
   Ⓒ 9
   Ⓓ None of these

10. Mrs. Williams went to Toys R' Us to purchase following items for each of her 3 children: one bicycle for $150, one bicycle helmet for $8, one arts and crafts set for $34 and one box of washable markers for $2. What is the total amount she spent before taxes?

    Ⓐ $194.00
    Ⓑ $582.00
    Ⓒ $572.00
    Ⓓ $482.00

Name _____   Date _____

Real World Problems (4.OA.A.2)

1. There are four boxes of pears. Each box has 24 pears. How many pears are there total?

    Ⓐ 72 pears
    Ⓑ 48 pears
    Ⓒ 96 pears
    Ⓓ 88 pears

2. Trevor has a collection of 450 baseball cards. He wants to place them into an album. He can fit 15 baseball cards on each page. How can Trevor figure out how many pages he will need to fit in all of his cards?

    Ⓐ By adding 450 and 15
    Ⓑ By subtracting 15 from 450
    Ⓒ By multiplying 450 by 15
    Ⓓ By dividing 450 by 15

3. Bow Wow Pet Shop has 12 dogs. Each dog had 4 puppies. How many puppies does the shop have in all?

    Ⓐ 16 puppies
    Ⓑ 12 puppies
    Ⓒ 48 puppies
    Ⓓ 36 puppies

4. Markers are sold in packs of 18 and 24. Yolanda bought five of the smaller packs and ten of the larger packs. How many markers did she buy altogether?

    Ⓐ 2 markers
    Ⓑ 432 markers
    Ⓒ 320 markers
    Ⓓ 330 markers

5. Each box of cookies contains 48 cookies. About how many cookies would be in 18 boxes?

    Ⓐ 100 cookies
    Ⓑ 500 cookies
    Ⓒ 1,000 cookies
    Ⓓ 2,000 cookies

6. At RTA Elementary School, there are 16 more female teachers than male teachers. If there are 60 female teachers, how can you find the number of male teachers in the school?

   Ⓐ Subtract 16 from 60
   Ⓑ Multiply 16 by 60
   Ⓒ Add 16 to 60
   Ⓓ Divide 60 by 16

7. Mrs. Willis wants to purchase enough Christmas ornaments so that each student can decorate 3 each. She has 24 students in her class. How many Christmas ornaments does she need to buy?

   Ⓐ 56
   Ⓑ 72
   Ⓒ 62
   Ⓓ 58

8. Jane needs to sell 69 cookie boxes for her scout troop. She has already sold 36 cookie boxes. How many more cookie boxes does she need to sell

   Ⓐ She needs to sell 36 more boxes.
   Ⓑ She needs to sell 39 more boxes.
   Ⓒ She needs to sell 33 more boxes.
   Ⓓ She needs to sell 69 more boxes.

9. RTA Elementary School had a 3-day food drive. On Monday, the school collected 10 soup cans and 16 cereal boxes. On Tuesday, the school collected 23 soup cans and 32 cereal boxes. On Wednesday, the school collected 36 soup cans and 44 cereal boxes. How many cereal boxes did the school collect in all?

   Ⓐ 76 cereal boxes
   Ⓑ 92 cereal boxes
   Ⓒ 161 cereal boxes
   Ⓓ 69 cereal boxes

10. There are 546 students in Hope School and 782 students in Trent School. How many more students are in Trent School than in Hope School?

    Ⓐ 136 students
    Ⓑ 236 students
    Ⓒ 144 students
    Ⓓ 244 students

Multi-Step Problems (4.OA.A.3)

1. Kristian purchased four textbooks which cost $34.99 each, and a backpack that cost $19.98. Estimate the total cost of the items he purchased. (You do not need to consider tax.)

    Ⓐ $90.00
    Ⓑ $160.00
    Ⓒ $120.00
    Ⓓ $55.00

2. During the last three games of the season, the attendance at the Tigers' home games was 14,667; 16,992; and 18,124. Estimate the total attendance for these three games. Round to the nearest thousand.

    Ⓐ 60,000
    Ⓑ 45,000
    Ⓒ 50,000
    Ⓓ 47,000

3. Steven keeps his baseball cards in an album. He has filled 147 pages of the album. He can fit 9 cards on each page. Which of the following statements is true?

    Ⓐ Steven has more than 2,000 baseball cards.
    Ⓑ Steven has between 1,000 and 1,500 baseball cards.
    Ⓒ Steven has between 1,500 and 2,000 baseball cards.
    Ⓓ Steven has less than 1,000 baseball cards.

4. Jam jars can be packed in large boxes of 60 or small boxes of 25. There are 700 jam jars to be shipped. The supplier wants to use the least number of boxes possible, but the boxes cannot be only partially filled. How many large boxes will the supplier end up using?

    Ⓐ 10 large boxes
    Ⓑ 11 large boxes
    Ⓒ 12 large boxes
    Ⓓ It is not possible to ship all 700 jars.

5. Allison needs 400 feet of rope to put a border around her yard. She can buy the rope in lengths of 36 feet. How many 36 foot long ropes will she need to buy?

    Ⓐ 9 ropes
    Ⓑ 10 ropes
    Ⓒ 11 ropes
    Ⓓ 12 ropes.

Name _____   Date _____

6. Katie and her friend went to the county fair. They each brought a $20.00 bill. The admission fee was $4.00 per person. Ride tickets cost 50 cents each. If each ride required two tickets per person, how many rides was each girl able to go on?

   Ⓐ 16 rides
   Ⓑ 32 rides
   Ⓒ 8 rides
   Ⓓ 20 rides

7. The population of the Bahamas is 276,208. The population of Barbados is 263,584. Which of the following statements is true of the total population of these two places?

   Ⓐ It is less than 500,000.
   Ⓑ It is between 500,000 and 550,000.
   Ⓒ It is between 550,000 and 600,000.
   Ⓓ It is more than 600,000.

8. Corey hopes to have 500 rocks in his collection by his next birthday. So far he has two boxes with 75 rocks each, a box with 85 rocks, and two boxes with 65 rocks each. How many more rocks does Corey need to gather to meet his goal?

   Ⓐ 145 rocks
   Ⓑ 135 rocks
   Ⓒ 275 rocks
   Ⓓ 235 rocks

9. Keith is helping his grandmother roll quarters to take to the bank. Each roll can hold 40 quarters. Keith's grandmother has told Keith that he can keep any leftover quarters, once the rolling is done. If there are 942 quarters to be rolled, how much money will Keith get to keep?

   Ⓐ $4.50
   Ⓑ $5.50
   Ⓒ $5.75
   Ⓓ $3.00

10. Assuming you are working with whole numbers, which of the following is not possible?

    Ⓐ Two numbers have a sum of 18 and a product of 72.
    Ⓑ Two numbers have a sum of 25 and a product of 100.
    Ⓒ Two numbers have a sum of 19 and a product of 96.
    Ⓓ Two numbers have a sum of 25 and a product of 144.

Name _____   Date _____

Number Theory (4.OA.A.4)

1. Andrew has a chart containing the numbers 1 through 100. He is going to put an "X" on all of the multiples of 10 and a circle around all of the multiples of 4. How many numbers will have an "X", but will not be circled?

   Ⓐ 3
   Ⓑ 4
   Ⓒ 5
   Ⓓ 8

2. Which number is a multiple of 30?

   Ⓐ 3
   Ⓑ 6
   Ⓒ 60
   Ⓓ 50

3. Use the Venn diagram below to respond to the following question.
   In which region of the diagram would the number 72 be found?

   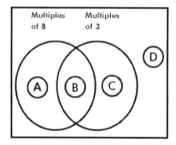

   Ⓐ Region A
   Ⓑ Region B
   Ⓒ Region C
   Ⓓ Region D

4. Which number can be divided evenly into 28?

   Ⓐ 3
   Ⓑ 6
   Ⓒ 7
   Ⓓ 5

5. Use the Venn diagram below to respond to the following question.
   Which of the following numbers would be found in Region D?

   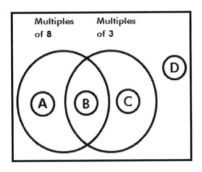

   Ⓐ 41
   Ⓑ 53
   Ⓒ 62
   Ⓓ All of the above

6. Which of these sets contains no composite numbers?

   Ⓐ 97, 71, 59, 29
   Ⓑ 256, 155, 75, 15
   Ⓒ 5, 23, 87, 91
   Ⓓ 2, 11, 19, 51

7. Choose the set that consists of only prime numbers.

   Ⓐ 2, 4, 8, 12
   Ⓑ 13, 15, 17, 19
   Ⓒ 2, 5, 23, 29
   Ⓓ 3, 17, 29, 81

8. Which of the following sets include factors of 44?

   Ⓐ 2, 4, 11, 22
   Ⓑ 0, 2, 6, 11
   Ⓒ 2, 4, 12, 22
   Ⓓ 4, 8, 12, 22

Name _____   Date _____

9. Which number completes the following number sentences?
   72 ÷ ___ = 6
   ___ x 6 = 72

   Ⓐ 6
   Ⓑ 8
   Ⓒ 12
   Ⓓ 16

10. If Carla wants to complete exactly 63 push ups during her workout, it is best if she does her push-ups in sets of _____.

    Ⓐ 13
    Ⓑ 8
    Ⓒ 7
    Ⓓ 17

End of Operations & Algebraic Thinking

Operations & Algebraic Thinking

Answer Key
&
Detailed Explanations

Number Sentences (4.OA.A.1)

Question No.	Answer	Detailed Explanation
1	D	It requires multiplication to find out the amount for twice as many. The symbol for multiplication is x. If **j** represents Josh's age, then **a** represents Andrew's age.
2	B	Mandy is making 4 equal groups out of 28. Therefore, 28 divided by 4 equal the number of marbles each friend receives.
3	B	To find n, we need to get it alone by subtracting the other numbers. This is an equation that needs to stay balanced, so what is done on one side of the = sign must be done on the other side. If we subtract 9 (6+3) from both sides, we have n = 13.
4	A	It is known that Cindy's mother baked 4 + 3 + 4 dozens of cookies plus an unknown number (n). The correct equation adds the amount baked Monday through Wednesday and adds the unknown (n).
5	B	47 divided by 9 = 5 with a remainder of 2.
6	D	There is a difference between the number of visitors to the science exhibit and the number of adult visitors. Subtract 63 from 147 to find n. The inverse equation is the correct answer: 63 + n = 147
7	B	89 divided by 11 is 8 with a remainder of 1. The remainder is the number of feet left over.
8	A	Adding 976 and 325 is the opposite of what the problem is stating: what number **subtracted** from 976 = 325.
9	C	Divide 486 by 54. 486 ÷ 54 = 9. Jack has 9 times as much money as Mary does.
10	B	For each child, Mrs. Williams spent $150 + 8 + 34 + 2 = $194.00. However the beginning of the problem states she is shopping for all three of her children so you will need to determine her full total. For three children, she would spend a total of $194.00 x 3 = $582.00.

Name _____ Date _____

Real World Problems (4.OA.A.2)

Question No.	Answer	Detailed Explanation
1	C	Each box has 24 pears and there are 4 boxes, so that is 24 x 4, which equals 96.
2	D	The total collection is 450. Each page will have 15 cards or parts of the whole. When you divide 15 into 450, the answer tells how many pages are needed.
3	C	That is 12 + 12 + 12 + 12, which is 12 x 4.
4	D	This problem requires 2 steps. There were (5) 18 count packs bought. Multiply 5 x 18 = 90. There were (10) 24 count packs bought. Multiply 10 x 24 = 240. Add the two products together to find out the sum or total amount bought. 240 + 90.
5	C	The question begins with the word, "about", which means to estimate. Begin by rounding off the number of cookies. 48 rounds off to 50. Multiply that estimate by the number of boxes: 50 x 18 = 900. Of the four estimate choices, 1,000 is the closest estimate.
6	A	Difference means subtraction and is the name of an answer to a subtraction problem. There are 60 female teachers, however, there are 16 more of them than male teachers: 16 less male teachers. Subtract 60 - 16 to find the answer.
7	B	Each student will decorate 3 ornaments. There are 24 students. That is 24, 3 times or 24 x 3.
8	C	This problem is asking the difference between how many cookies need to be sold and how many are already sold. Therefore, 69 - 36=33.
9	B	The question only asks for the total number of cereal boxes. Everything pertaining to soup cans is extra, unnecessary information. Add 16 + 32 + 44.
10	B	To find the difference between how many more students are at Trent School than Hope school, subtract the amounts: 782 - 546.

Multi-Step Problems (4.OA.A.3)

Question No.	Answer	Detailed Explanation
1	B	Round off the cost of the textbooks to the nearest dollar. $34.99 is close to $35.00. Round off the price of the backpack from $19.98 to $20.00. The estimated amount Kristian paid can be found be calculating 4 x $35.00 + $20.00. 4 x $35.00 = $140.00; $140.00 + 20.00= $160.00
2	C	To find the estimated total, round each number to the nearest thousand, then add. 14,667 rounds to 15,000; 16,992 rounds to 17,000; and 18,124 rounds to 18,000. 15,000 + 17,000 + 18,000 = 50,000
3	B	To decide which choice is true, use estimation. Steven has filled about 150 pages in his album. (147 is close to 150.) There are about 10 cards on each page. Therefore, he has about 1,500 cards. (150 x 10 = 1,500) Since both of the numbers were rounded up, this estimate is an overestimate. Steven, therefore, has slightly less than 1,500 cards. The second choice is the most reasonable.
4	A	The best option for the supplier would be to use as many large boxes as possible. The supplier would be able to ship 600 jars in 10 large boxes and 100 jars in 4 small boxes. The supplier could not use any more large boxes, because 11 large boxes would contain 660 jars. That would leave 40 more jars to be shipped. That cannot be done without sending a partially-filled box.
5	D	If Allison were to buy 10 ropes, she would have 360 feet of rope. (36 x 10 = 360) Buying one more rope would bring her total up to 396 feet. (360 + 36 = 396) She is still 4 feet short. Therefore, Allison will have to buy 12 ropes in order to get at least 400 feet of rope.
6	A	Because the problem is asking about the number of rides **each** girl could go on, you can focus just on Katie. Katie brought $20.00 with her to the fair. After paying the $4.00 admission fee, she has $16.00 left over. If each ride required two tickets, and the tickets are each worth 50 cents each, then each ride costs $1.00 per person. With $16.00 left over, Katie could go on 16 rides.
7	B	An estimated sum could be found by using compatible numbers. 276,208 is close to 275,000; 263,584 is close to 265,000. 275,000 + 265,000 = 540,000, which is between 500,000 and 550,000.

Question No.	Answer	Detailed Explanation
8	B	First, total up the rocks that Corey already has in his collection: (2 x 75) + (2 x 65) + 85 = 150 + 130 + 85 = 365. Then subtract what he has from 500: 500 - 365 = 135.
9	B	Divide the total number of quarters by 40 to see how many rolls can be made. The remainder will be the quarters that Keith gets to keep. 942 ÷ 40 = 23 R 22. There can be 23 rolls, with 22 quarters left over. Keith will keep the 22 quarters. The value of 22 quarters (4 quarters make a dollar) is $5.50.
10	C	Solution for choice A: 12 and 6 Solution for choice B: 20 and 5 Solution for choice D: 16 and 9 There is no solution for choice C.

Number Theory (4.OA.A.4)

1	C	Multiples are the products of two numbers. Skip count, recite time-tables or refer to a chart to find all of the multiples of both numbers, from 1 to 100. Listing these multiples may also be helpful: 10 is 10, 20, 30, 40, 50, 60, 70, etc. 4 is 4, 8, 12, 16, 20, 24, 28, 32, etc. All of the multiples of 10 will have an x and all of multiples of 4 will be circled. The question to the problem is **how many numbers will have an x, but *not* be circled.** In the list of 10 multiples, circle all the multiples of 4. Count the number of multiples of 10 that do not have a circle.
2	C	The multiple of 30 must be a product of 30 x another number. 60 = 30 x 2.
3	B	In this diagram, Section A would contain the multiples of 8 which are not multiples of 3. Section C would contain the multiples of 3 which are not multiples of 8. Section B would contain multiples of both 8 and 3. Section D would list numbers that are **not** multiples of 8 or 3. 8 x 9 = 72 and 3 x 24 = 72, so 72 would be found in Section B.
4	C	Using the inverse relationship between multiplication and division to choose the number that is a factor of 28. 28 is a multiple of 7, so 7 is a factor of 28.
5	D	Refer to the lists of the multiples of 8 and 3. The D section of the Venn diagram is for numbers that are not multiples of 8 or 3. 41, 53, and 62 are not multiples of 3 or 8.

Name _____ Date _____

Question No.	Answer	Detailed Explanation
6	A	Any whole number greater than 1 is either classified as composite or prime. Therefore, this question is asking you to find the set that contains only prime numbers. A prime number is a number that has only 2 factors - itself and 1. The first set of numbers fits this criteria. All four of the numbers in the set have only two factors.
7	C	Prime numbers are numbers greater than 1 that have only two factors: themselves and the number 1.
8	A	Choose the set that gives the product of 44 when multiplied with another number in that set.
9	C	12 x 6 = 72, so 72 ÷ 12 = 6
10	C	63 is a multiple of 7. 63 is not a multiple of 13, 8, or 17. In order to do exactly 63 push-ups, Carla is best to do sets of 7.

Name _____   Date _____

Number & Operations in Base Ten

Section 1: Generalize place value understanding for multi-digit whole numbers.

Place Value (4.NBT.A.1)

1. What number can be found in the ten-thousands digit of 291,807?

   Ⓐ 9
   Ⓑ 1
   Ⓒ 2
   Ⓓ 0

2. Consider the number 890,260.
   The 8 is found in the _____ place.

   Ⓐ ten-thousands
   Ⓑ millions
   Ⓒ thousands
   Ⓓ hundred-thousands

Place Value Chart

Hundred-billions	Ten-billions	Billions	Hundred-millions	Ten-millions	Millions	Hundred-thousands	Ten-thousands	Thousands	Hundreds	Tens	Ones

Name _____   Date _____

3. What number correctly completes this statement?
   9 ten thousands = _____ thousands

   Ⓐ 90
   Ⓑ 900
   Ⓒ 9
   Ⓓ 19

4. Which number is in the thousands place in the number 984,923?

   Ⓐ 9
   Ⓑ 8
   Ⓒ 4
   Ⓓ 2

5. What is the value of the 8 in 683,345?

   Ⓐ 80
   Ⓑ 800
   Ⓒ 8,000
   Ⓓ 80,000

6. Which number equals 4 thousands, 6 hundreds, 0 tens, and 5 ones?

   Ⓐ 465
   Ⓑ 4,605
   Ⓒ 4,650
   Ⓓ 4,065

7. What number is in the tens place in 156.25?

   Ⓐ 1
   Ⓑ 5
   Ⓒ 6
   Ⓓ 2

8. Which number equals 2 ten thousands, 1 hundred thousand, and 3 ones

   Ⓐ 120,003
   Ⓑ 210,003
   Ⓒ 102,003
   Ⓓ 213,000

Name _____ Date _____

9. Which answer shows the value of each 7 in this number: 7,777?

   Ⓐ 7,000; 700, 70, 7
   Ⓑ 7 x 7 x 7 x 7
   Ⓒ 700,000, 70,000, 700, 70
   Ⓓ 7 + 7 + 7 + 7

10. Mrs. Winters went to the bank with eight 100 dollar bills. She wanted to replace them with all 10 dollar bills. How many 10 dollar bills will the bank give her in exchange?

    Ⓐ 800 ten dollar bills
    Ⓑ 8,000 ten dollar bills
    Ⓒ 8 ten dollar bills
    Ⓓ 80 ten dollar bills

Name _____   Date _____

Compare Numbers and Expanded Notation (4.NBT.A.2)

1. Arrange the following numbers in ascending order.
   62,894; 26,894; 26,849; 62,984

   Ⓐ 62,984; 62,894; 26,894; 26,849
   Ⓑ 26,894; 26,849; 62,984; 62,894
   Ⓒ 26,849; 62,984; 62,894; 26,894
   Ⓓ 26,849; 26,894; 62,894; 62,984

2. Which of the following statements is true?

   Ⓐ 189,624 > 189,898
   Ⓑ 189,624 > 189,246
   Ⓒ 189,624 < 189,264
   Ⓓ 189,624 = 189,462

3. Which number will make this statement true?
   198,888 > _____

   Ⓐ 198,898
   Ⓑ 198,879
   Ⓒ 198,889
   Ⓓ 199,888

4. Which statement is NOT true?

   Ⓐ 798 < 799
   Ⓑ 798 > 789
   Ⓒ 798 < 789
   Ⓓ 798 = 798

5. Write the expanded form of this number.
   954,351

   Ⓐ 90,000 + 5,000 + 400 + 30 + 5 + 1
   Ⓑ 900,000 + 50,000 + 4,000 + 300 + 50 + 1
   Ⓒ 900,000 + 54,000 + 300 + 50 + 1
   Ⓓ 900,000 + 50,000 + 4,000 + 300 + 51

26

© Lumos Information Services 2015 | LumosLearning.com

Name _____   Date _____

6. What is the standard form of this number?
   30,000 + 200 + 50

   Ⓐ 30,250
   Ⓑ 32,500
   Ⓒ 325,000
   Ⓓ 3,250

7. Write the standard form of:
   8 ten thousands, 4 thousands, 1 hundred, 6 ones

   Ⓐ 84,160
   Ⓑ 84,106
   Ⓒ 8,416
   Ⓓ 84,016

8. Write 1,975,206 in expanded form.

   Ⓐ 1,000,000 + 9,000,000 + 7,000 + 500 + 20 + 6
   Ⓑ 1,000,000 + 9,000,000 + 70,000 + 5,000 + 200 + 60
   Ⓒ 100,000 + 900,000 + 70,000 + 5,000 + 200 + 6
   Ⓓ 1,000,000 + 900,000 + 70,000 + 5,000 + 200 + 6

9. What is 300,000 + 40,000 + 20 + 5 in standard form?

   Ⓐ 34,025
   Ⓑ 3,425
   Ⓒ 340,025
   Ⓓ 342,005

10. Write the standard form for 2 hundred thousands, 1 ten thousand, 4 hundreds, 1 ten, 9 ones.

    Ⓐ 210,419
    Ⓑ 201,419
    Ⓒ 200,419
    Ⓓ 21,419

Name _____   Date _____

Rounding Numbers (4.NBT.A.3)

1. Round 4,170,154 to the nearest hundred.

   Ⓐ 4,200,000
   Ⓑ 4,170,100
   Ⓒ 4,170,000
   Ⓓ 4,170,200

2. Round 4,170,154 to the nearest thousand.

   Ⓐ 4,200,000
   Ⓑ 180,000
   Ⓒ 4,170,000
   Ⓓ 4,179,200

3. Round 4,170,154 to the nearest ten thousand.

   Ⓐ 4,200,000
   Ⓑ 4,170,000
   Ⓒ 4,179,000
   Ⓓ 4,179,200

4. Round 4,170,154 to the nearest hundred thousand.

   Ⓐ 4,200,000
   Ⓑ 4,180,000
   Ⓒ 4,179,000
   Ⓓ 4,100,000

5. Round 4,170,154 to the nearest million.

   Ⓐ 4,000,000
   Ⓑ 4,180,000
   Ⓒ 4,179,000
   Ⓓ 5,000,000

Name _____ Date _____

6. Round 424,819 to the nearest ten.

   Ⓐ 400,820
   Ⓑ 424,810
   Ⓒ 424,020
   Ⓓ 424,820

7. The soda factory bottles 2,451 grape and 3,092 orange sodas each day. About how many of the two types of soda are bottled each day? Round to the nearest hundred.

   Ⓐ 5,400 sodas
   Ⓑ 5,000 sodas
   Ⓒ 5,600 sodas
   Ⓓ 5,500 sodas

8. The Campbells have 3,000 books in their personal library. If 1,479 of the books are fiction, and the rest are non-fiction, about how many are non-fiction? Round to the nearest hundred.

   Ⓐ 2,500 books
   Ⓑ 1,400 books
   Ⓒ 1,500 books
   Ⓓ 1,600 books

9. Solve; Give an estimate as the final answer.
   $$\begin{array}{r} 7000 \\ -4258 \\ \hline \end{array}$$

   Ⓐ 2,742
   Ⓑ 3,000
   Ⓒ 2,000
   Ⓓ 6,858

10. The school has $925 to spend on new books for the library. Each book costs $9.95. Estimate how many books can be bought.

    Ⓐ 70
    Ⓑ 80
    Ⓒ 90
    Ⓓ 120

Name _____   Date _____

Section 2: Use place value understanding and properties of operations to perform multi-digit arithmetic.

Addition & Subtraction (4.NBT.B.4)

1. What number acts as the identity element in addition?

   Ⓐ 2-1
   Ⓑ 0
   Ⓒ 1
   Ⓓ None of these

2. Which of the following number sentences illustrates the Commutative Property of Addition?

   Ⓐ 3 + 7 = 7 + 3
   Ⓑ 9 + 4 = 10 + 3
   Ⓒ 11 + 0 = 11
   Ⓓ 2 + (3 + 4) = 2 + 7

3. What number makes this number sentence true?
   10 + ___ = 0

   Ⓐ 10
   Ⓑ 1/10
   Ⓒ 0
   Ⓓ -10

4. Find the sum.
   24 + 37 + 76 + 13

   Ⓐ 140
   Ⓑ 150
   Ⓒ 151
   Ⓓ none of these

Name _____ Date _____

5. Find the difference.
   702 - 314 = _____

   Ⓐ 388
   Ⓑ 412
   Ⓒ 312
   Ⓓ 402

6. What is the sum of 0.55 + 6.35?

   Ⓐ 6.09
   Ⓑ .85
   Ⓒ 6.90
   Ⓓ none of these

7. Find the difference.
   7.86 - 4.88

   Ⓐ 2.98
   Ⓑ 3.02
   Ⓒ 2.08
   Ⓓ 1.98

8. Find the sum.
   156 + 99 =

   Ⓐ 256
   Ⓑ 245
   Ⓒ 265
   Ⓓ 255

9. There are 1,565 pictures on the disks. Only 1,430 of them are in color. How many pictures are not color?

   Ⓐ 135 pictures
   Ⓑ 2,995 pictures
   Ⓒ 2,195 pictures
   Ⓓ 995 pictures

Name _____   Date _____

10. James got 300 coins while diving in the game, The Amazing World of Gumball Splashmasters! However, he hit 2 birds and lost 12 coins. During the second round, he got 250 coins and hit no birds. How many coins did he have at the end of the second round?

   Ⓐ 500 coins
   Ⓑ 542 coins
   Ⓒ 548 coins
   Ⓓ 538 coins

Name _____   Date _____

Multiplication (4.NBT.B.5)

1. Assume a function table has the rule "Multiply by 6." What would the OUT value be if the IN value was 8?

   Ⓐ 14
   Ⓑ 48
   Ⓒ 16
   Ⓓ 32

2. Solve.
   26 x 8 = ____

   Ⓐ 206
   Ⓑ 168
   Ⓒ 182
   Ⓓ 208

3. The number sentence 4 x 1 = 4 illustrates which mathematical property?

   Ⓐ The Associative Property of Multiplication
   Ⓑ The Identity Property of Multiplication
   Ⓒ The Distributive Property
   Ⓓ The Associative Property of Multiplication

4. Find the product of 17 x 6.

   Ⓐ 102
   Ⓑ 84
   Ⓒ 119
   Ⓓ 153

Name _____ Date _____

5. If the IN value is 0, what is the OUT value?

   RULE: Divide by 9

IN	OUT
0	??
9	??
81	9
99	11

   Ⓐ 2
   Ⓑ 0
   Ⓒ 9
   Ⓓ 1

6. Which of the following number sentences illustrates the Associative Property of Multiplication?

   Ⓐ 4 x 0 = 0
   Ⓑ 77 x 1 = 1 x 77
   Ⓒ (2 x 4) x 5 = 2 x (4 x 5)
   Ⓓ 13 x 7 = (10 x 7) + (3 x 7)

7. Solve.
   4 x 3 x 6 = _____

   Ⓐ 48
   Ⓑ 72
   Ⓒ 56
   Ⓓ 64

8. Find the exact product of 5 x 20 x 8.

   Ⓐ 800
   Ⓑ 560
   Ⓒ 80
   Ⓓ 900

9. Complete the following statement:
   "In a multiplication sentence, if a factor is 1, then _____"

   Ⓐ the other factor is also 1
   Ⓑ the product is also 1
   Ⓒ the other factor and the product are the same
   Ⓓ the product is 0

10. Steve has 7 pages in his stamp collection book. Each page holds 20 stamps. How many total stamps does Steve have in his collection?

    Ⓐ 14
    Ⓑ 6
    Ⓒ 140
    Ⓓ 8

Name _____ Date _____

Division (4.NBT.B.6)

1. What role does the number 75 play in the following equation?
   300 ÷ 75 = 4

   Ⓐ It is the dividend.
   Ⓑ It is the quotient.
   Ⓒ It is the divisor.
   Ⓓ It is the remainder.

2. Which of the following division expressions will have no remainder?

   Ⓐ 73 ÷ 9
   Ⓑ 82 ÷ 6
   Ⓒ 91 ÷ 7
   Ⓓ 39 ÷ 9

3. Divide these blocks into 2 equal groups. How many will be in each group?
   Note: 1 flat = 10 rods. 1 rod = 10 cubes

      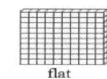
   flat   flat   flat   rods   cubes

   Ⓐ 352
   Ⓑ 176
   Ⓒ 152
   Ⓓ 132

4. Fill in the blank:
   480 ÷ 6 = (400 ÷ 6) + (80 ÷ ___)

   Ⓐ 400
   Ⓑ 6
   Ⓒ 80
   Ⓓ 480

5. What amount would be in each group if this number were divided into 6 groups?
Note: 1 flat = 10 rods. 1 rod = 10 cubes

flat

flat

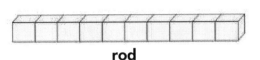
rod

Ⓐ 35
Ⓑ 45
Ⓒ 25
Ⓓ 15

6. Find the quotient:
694 ÷ 2 = ____

Ⓐ 342
Ⓑ 327
Ⓒ 347
Ⓓ 332

7. What is the remainder when 100 is divided by 12?

Ⓐ 4
Ⓑ 6
Ⓒ 8
Ⓓ 2

8. 28 x 6 = 168
Find a related fact to the one shown above.

Ⓐ 28 ÷ 16 = 8
Ⓑ 168 ÷ 6 = 28
Ⓒ 168 ÷ 28 = 8
Ⓓ 168 ÷ 8 = 16

9. The Thomas' took 96 bottles of spring water to the family reunion picnic. If they had purchased 4 identical cases, how many bottles were in each case?

Ⓐ 12
Ⓑ 48
Ⓒ 24
Ⓓ 18

Name _____ Date _____

10. Each of Mrs. Harris' 7 children collected the same number of exotic insects for their individual displays. How many insects did each child have for show and tell, if they had a total of 112 insects?

Ⓐ 112
Ⓑ 16
Ⓒ 7
Ⓓ 42

End of Number & Operations in Base Ten

Name _____ Date _____

Number and Operations in Base Ten

Answer Key
&
Detailed Explanations

Name _____   Date _____

Place Value (4.NBT.A.1)

Question No.	Answer	Detailed Explanation
1	A	Place values are read from right to left, beginning with the "ones" place, "tens", "hundreds", "thousands", "ten thousands", "hundred thousands", "millions", etc. If you were to write the number in the boxes below, you see the 9 is in the ten-thousand column. *[Place Value Chart showing: Hundred-billions, Ten-billions, Billions, Hundred-millions, Ten-millions, Millions, Hundred-thousands, Ten-thousands, Thousands, Hundreds, Tens, Ones]*
2	D	Begin naming the place values for each number from the right. Number 9 is in the ten thousands place. Place values increase by multiplying 10: 1 ten is 10, 10 tens is a hundred, 10 hundreds is a thousand, etc.
3	A	Multiply 9 x 10,000 to find 90,000.
4	C	Number 3 is in the "ones" place. Number 2 is in the "tens" place. Number 9 is the "hundreds" place. The "thousands" place is next.
5	D	The 8 is in the ten thousands place, which is 8 x 10,000.
6	B	Write the 4 in the thousands place, the 6 in the hundreds place, the 0 in the tens place and the 5 in the ones place.
7	B	Numbers to the right of the decimal point begin with the value of tenths, hundredths, etc. Numbers to the left of the decimal place are the ones, tens, hundreds, etc. *[Diagram showing 9,605,872.14573 with labels: one millions, hundred thousands, ten thousands, one thousands, hundreds, tens, ones, tenths, hundredths, thousandths, ten thousandths, hundred thousandths, millionths]*

40

Question No.	Answer	Detailed Explanation
8	A	Though not stated as such in the problem, the digit in the hundred thousands place is written first. The 2 ten thousands is written next: 2 ten thousands is 2 x 10,000. The next place that has any value is the ones place, which has 3. The thousands and hundreds place have no value, so zeros are placed there.
9	A	Write the numbers in expanded notation, which shows the entire value of the number written out. 7 in the thousands place is written as 7,000. 7 in the hundreds place is written as 700. 7 in the tens place is written as 70. 7 in the ones place is written as 7.
10	D	There are ten 10 dollar bills in $100. Therefore, there are 80 ten-dollar bills in $800.

Compare Numbers and Expanded Notation (4.NBT.A.2)

1	D	Ascending order is from least to greatest value. All of these numbers start in the ten thousands place, however the numbers beginning with a 2 have a lesser value. If the numbers are the same in the thousands place, for example, compare the numbers that are different in the hundreds place. If the digits are the same in the hundreds place, compare the tens.
2	B	6 hundreds, 2 tens and 4 ones is greater than 2 hundreds, 6 tens and 4 ones. The numbers are the same in all places above the ten places so you have to look there.
3	B	8 x 10 is less than 9 x 10 and greater than 7 x 10.
4	C	Compare all of the digits in the numbers: 9 x 10 is not < 8 x 10.
5	B	This form requires that every digit be multiplied by the multiple of ten that corresponds to its place value and then written as such. For example, the 9 is in the hundred thousands place, which is 9 x 100,000 and written as 900,000. Place a plus sign (+) between each term in the expanded form.
6	A	There are no thousands and no ones in this number. Place a 0 in those positions.

Question No.	Answer	Detailed Explanation
7	B	The standard form is simply the digits in the number placed into their appropriate places. A zero will be placed in the tens place, since there are no tens listed in the problem.
8	D	The standard form is a 7-digit number, so the expanded form must show numbers in the millions place. There are no tens in this number, so in expanded form, this place value is simply omitted.
9	C	Zeroes should be written in the thousands and hundreds places. All other digits will be placed according to their corresponding place values.
10	A	There are no thousands in this number, so a 0 will be placed in the thousands place. All other digits are placed according to their corresponding place values.

Rounding Numbers (4.NBT.A.3)

1	D	Since this number is to be rounded to the nearest hundred, the number to the right of the hundreds place (the tens place) will determine if the hundreds digit is to be rounded up to the next number or stay the same. If the number in the tens place is 0-4, keep the hundreds place the same. If the number in the tens place is between 5 and 9 (inclusive), round the hundred up to the next number. Example 678 would round up to 700; whereas 628 would round down to 600.
2	C	The number in the hundreds place is a 1, so the thousands place will not change. It will remain a zero. The hundreds, tens, and ones places will also become zeroes.
3	B	Use the digit in the thousands place to decide how to round. In this number, the thousands place has a zero, so the ten thousands place will remain unchanged when rounding.
4	A	Since the digit in the ten thousands place is 5 or greater (it is a 7), round up to the next hundred thousand.
5	A	Round this number based on the digit that is in the hundred thousands place. If it is a 0-4, the digit in the millions place will remain the same. If the digit in the hundred thousands place is a 5-9, the digit in the millions place will round up to the next digit.

42

© Lumos Information Services 2015 | LumosLearning.com

Question No.	Answer	Detailed Explanation
6	D	The digit 9 in the ones place will cause the number to round up to the next ten. When rounding to the nearest ten, the digit in the ones place always becomes a zero.
7	C	First round the two values given in the problem to the nearest hundred. 2,451 rounds to 2,500. 3,092 rounds to 3,100. Then add the rounded values to find the estimated sum. 2,500 + 3,100 = 5,600
8	C	First round each of the values given in the problem to the nearest hundred. 3,000 rounds to 3,000. 1,479 rounds to 1,500 Then subtract the rounded values to find the estimated difference. 3,000 - 1,500 = 1,500
9	B	Regrouping cannot be done with a zero until it has been renamed and has a different value.
10	C	Using reasoning to estimate the amount of books that could be purchased: Each book costs about $10.00. The library has around $900.00 to spend. Dividing 900 by 10, yields a quotient of 90. The school can purchase about 90 books for its library.

Addition & Subtraction (4.NBT.B.4)

1	B	Any number plus 0 is **always** that number, so 0 is the additive identity.
2	A	The Commutative Property means that changing the order of the addends (the two numbers being added together) does not change the sum.
3	D	Adding a negative number is the same as subtracting it. For example, 6 + -6 is the same as 6 - 6. Also, 10 and -10 are additive inverses, so their sum must be the additive identity, 0.
4	B	The same rules apply when adding more than two numbers: 24 37 76 13

Name _____   Date _____

Question No.	Answer	Detailed Explanation
5	A	Line numbers up. The first one is written down and the second number is written underneath. Subtract the number on the bottom from the number on the top. If the number on the top is smaller than the one on the bottom, regroup: you cannot subtract 4 from 2, so take one from the tens place. Oops! That is a zero, so take 1 from the hundreds place. That leaves a 6 in the hundreds place. Bring that 1 to the tens place and then borrow it again. Take it over to the ones place where it is needed. That leaves a 9 in the tens place and the 2 becomes a 12. Now subtract: 702 <u>314</u>
6	C	Numbers with decimal points are added just like whole numbers, except the decimal points must be in alignment directly underneath each other, including the answer: 0.55 <u>6.35</u>
7	A	Eight cannot be taken from 6, so regroup in the tens place. That 8 becomes a 7 and the 6 in the ones place becomes a 16. Now subtract, regrouping again in the hundreds place so that 8 can be subtracted from 17. The 7 in the hundreds place becomes a 6. Plot a decimal point in the answer directly beneath the other decimal points.
8	D	You can round 99 to 100, and then add 156. Then subtract 1.
9	A	Subtract the number of pictures that are color from the total number of pictures: 1,565 - 1,430. Regrouping is not necessary in this problem.
10	D	Add up the total number of coins won in the first game, then subtract the 12 coins lost. Then add up the number of coins won in the second game. Add the total coins won from both games together.

Multiplication (4.NBT.B.5)

1	B	Apply the rule to each IN value: 8 x 6
2	D	This problem could be solved using the Distributive Property: 26 x 8 = (20 + 6) x 8 = (20 x 8) + (6 x 8) = 160 + 48 = 208
3	B	Any number times 1 will **always** equal that number.
4	A	This problem could be solved using the Distributive Property: 17 x 6 = (10 + 7) x 6 = (10 x 6) + (7 x 6) = 60 + 42 = 102

Name _____   Date _____

Question No.	Answer	Detailed Explanation
5	B	The rule in this table is to divide the IN number by 9. $0 \div 9 = 0$.
6	C	The order in which numbers are grouped and multiplied does not change the product.
7	B	Even without parenthesis, the order in which the numbers are multiplied will not change the product: Multiply 4 x 3. Then multiply that product x 6. $4 \times 3 = 12$ and $12 \times 6 = 72$.
8	A	This can be solved using mental math. Multiply $5 \times 20 = 100$. Then multiply $100 \times 8 = 800$.
9	C	Any number multiplied by 1 will **always** have that number as the product, which is the Identity Property of Multiplication.
10	C	One page holds 20 stamps. Steve has 7 pages, therefore total stamps with Steve, $7 \times 20 = 140$.

Division (4.NBT.B.6)

1	C	The number that follows the division symbol ($\div$) is called the divisor.
2	C	$91 \div 7$ will have no remainder, since 91 is a multiple of 7. $7 \times 13 = 91$.
3	B	Count the value of the base 10 blocks that are shown, which is 352. Because there is only one set of blocks, regroup each hundreds flat into 10 tens rods. Regroup each tens rod into ten ones cubes. Place 1 tens rod and 1 ones cube into 2 groups until all are placed. Count the value of the tens blocks shown in both groups. The numbers should be the same.
4	B	Both numbers that make up the dividend (480) must be divided by the divisor (6).
5	A	The number being modeled is 210. Regroup all of the hundreds flats into tens rods. Place 1 tens rod and 1 ones cube into 6 groups until all are placed, regrouping tens for ones when necessary. Count by tens and ones to determine the value amount of blocks in each group. That number should be the same in each group and is the answer.
6	C	Begin by dividing 6 by 2, following these steps in this order: 1. Divide 2. Multiply 3. Subtract 4. Bring down the next number in the dividend. Multiply the 2 by 3, which is the number of times 2 divides into 6. Place the 3 above the 6. $2 \times 3 = 6$. Place that product under the 6 in the dividend. Subtract the two 6's. Bring down the next number in the dividend, which is 9 and write it next to the 0. Repeat the four steps for each number in the dividend.

© Lumos Information Services 2015 | LumosLearning.com

Name _____   Date _____

Question No.	Answer	Detailed Explanation
7	A	12 x 8 = 96, so 100 ÷ 12 would have a remainder of 4, since 100 is 4 greater than 96.
8	B	The product in the multiplication problem is the dividend in the division problem. One factor becomes the divisor and the other one becomes the quotient. Therefore, 168 ÷ 6 = 28 is related to 28 x 6 = 168.
9	C	Divide 96 by 4. 96 ÷ 4 = (100 - 4) ÷ 4 = (100 ÷ 4) - (4 ÷ 4) = 25 - 1 = 24
10	B	Divide 112 by 7. 112 ÷ 7 = (70 + 42) ÷ 7 = (70 ÷ 7) + (42 ÷ 7) = 10 + 6 = 16

Name _____   Date _____

Number & Operations – Fractions

Section 1: Extend understanding of fraction equivalence and ordering.

Equivalent Fractions (4.NF.A.1)

1. What fraction of these shapes are squares?

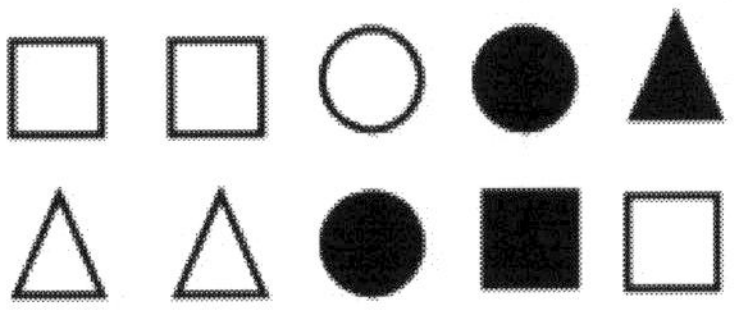

   ▢ ▢ ○ ● ▲
   △ △ ● ▪ ▢

   Ⓐ 1/4
   Ⓑ 4/6
   Ⓒ 4/10
   Ⓓ 1/3

2. What fraction of these shapes are not circles?

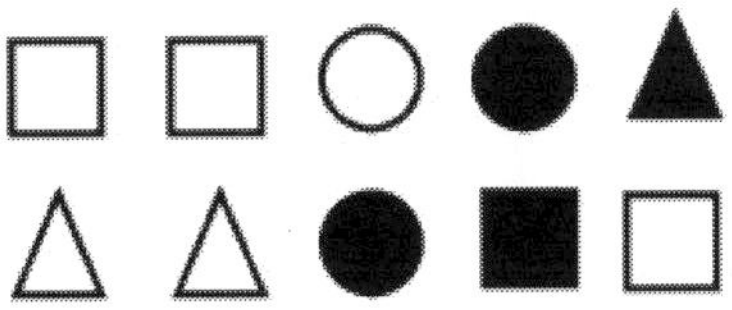

   ▢ ▢ ○ ● ▲
   △ △ ● ▪ ▢

   Ⓐ 3/7
   Ⓑ 8/10
   Ⓒ 7/10
   Ⓓ 1/3

Name _____ Date _____

3. What fraction of the squares are shaded?

Ⓐ 1/4
Ⓑ 1/10
Ⓒ 1/3
Ⓓ 3/4

4. What fraction of the shaded shapes are circles?

Ⓐ 2/10
Ⓑ 1/3
Ⓒ 2/2
Ⓓ 2/4

5. Continue the pattern of equivalent fractions:
   1/2, 2/4, 3/6, 4/8...
   What fraction would come next in the pattern?

Ⓐ 1/3
Ⓑ 1/16
Ⓒ 5/10
Ⓓ 3/4

6. Which pair of addends have the fraction 11/12 as a sum?

Ⓐ 9/6 + 2/6
Ⓑ 7/12 + 4/12
Ⓒ 9/12 + 1/12
Ⓓ 11/12 + 1/1

7. Which fraction is equivalent to this model?

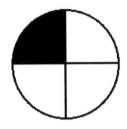

Ⓐ 1/5
Ⓑ 3/7
Ⓒ 2/7
Ⓓ 4/16

8. Which fraction is equivalent to 8/18?

Ⓐ 1/5
Ⓑ 3/7
Ⓒ 2/7
Ⓓ 4/9

9. Continue the pattern of equivalent fractions:
5/6, 10/12, 15/18,...
What fraction would come next in the pattern?

Ⓐ 7/14
Ⓑ 20/24
Ⓒ 9/45
Ⓓ 12/36

10. Reduce the fraction 21/49 to its lowest terms:

Ⓐ 1/5
Ⓑ 3/7
Ⓒ 2/7
Ⓓ 4/9

11. Reduce the fraction 44/49 to its lowest terms:

    Ⓐ 1/5
    Ⓑ 3/7
    Ⓒ 2/7
    Ⓓ 4/9

12. Patrick climbed 4/5 of the way up the trunk of a tree. Jacob climbed 80/100 of the way up the same tree. To accomplish the same distance as Patrick and Jacob, how far up that tree trunk will Devon have to climb?

    Ⓐ 15/20
    Ⓑ 60/75
    Ⓒ 100/200
    Ⓓ 28/42

13. The cheerleaders ate 9/18 of a sheet cake. Write this fraction in lowest terms.

    Ⓐ 1/9
    Ⓑ 1/2
    Ⓒ 2/3
    Ⓓ 3/6

14. Which group of fractions can all be reduced to 2/9?

    Ⓐ 23/27, 4/36, 30/270
    Ⓑ 25/50, 30/60, 50/100
    Ⓒ 4/18, 6/27, 50/225
    Ⓓ 6/21, 20/70, 36/84

15. What do these fractions have in common?
    10/16; 15/24; 20/32; 25/40; 30/48

    Ⓐ These fractions are equivalent to 5/9.
    Ⓑ These fractions are equivalent to 5/8.
    Ⓒ These fractions are equivalent to 10/12.
    Ⓓ These fractions are equivalent to 4/8.

Compare Fractions (4.NF.A.2)

1. Where is Point D located on this number line?

   A) -2.5
   B) -2
   C) -1.5
   D) -3

2. Which statement is true?

   A) 4/14=6/21=8/28
   B) 4/14>6/21>8/28
   C) 4/14<6/21<8/28
   D) 4/14<6/21>8/28

3. Compare the two fractions using < = or >:

   3/12 ———— 3/18

   A) =
   B) <
   C) >

4. Compare the two fractions using < = or >:

   4/28 ———— 4/20

   A) =
   B) <
   C) >

Name _____   Date _____

5. Which symbol makes this statement true?
   4/9 + 3/9 ___ 6/9

   Ⓐ >
   Ⓑ =
   Ⓒ <

6. Which symbol makes this statement true?
   75/100 - 32/100 ___ 42/100

   Ⓐ >
   Ⓑ =
   Ⓒ <

7. Which fraction below has a greater value than the fraction being shown?

   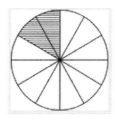

   Ⓐ 2/20
   Ⓑ 2/15
   Ⓒ 2/10
   Ⓓ 2/100

8. The popsicle slowly melted in the hot sun. Which group of fractions could represent the amount of popsicle remaining after 2 minutes, 4 minutes, and 6 minutes had passed?

   Ⓐ 1/3, 1/2, 3/4
   Ⓑ 3/4, 1/2, 1/3
   Ⓒ 1/3, 3/4, 1/2
   Ⓓ 3/4, 1/3, 1/2

9. Arrange these models in order from greatest to least:

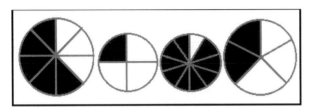

Model A   Model B   Model C   Model D

- Ⓐ A, B, C, D
- Ⓑ C, A, D, B
- Ⓒ C, D, B, A
- Ⓓ A, C, D, B

10. Which sum is greater?
    692/1000 + 231/1000 ___ or 245/1000 + 726/1000

- Ⓐ <
- Ⓑ =
- Ⓒ >

11. These fractions are arranged from least to greatest. Which fraction could go in the blank?
    9/25, ___, 16/25, 21/25

- Ⓐ 13/25
- Ⓑ 6/25
- Ⓒ 18/25
- Ⓓ 23/25

12. Marty eats vegetables 6/7 days out of the week. Dan eats them 3/7 days out of the week. How many more days does Marty eat vegetables each week?

- Ⓐ 2 days
- Ⓑ 1 day
- Ⓒ 3 days
- Ⓓ 4 days

13. The salesman sold 1/3 of his inventory during a weekend sale. He had hoped to sell an even higher amount. Which fraction could represent the amount of his inventory he had hoped to sell?

   Ⓐ 1/4
   Ⓑ 1/8
   Ⓒ 1/2
   Ⓓ 1/6

14. One fifth of the tourists went to see the natural waterfall on Monday. Five fifths of them went to see it on Tuesday, and three fifth of them went to see it on Wednesday. List the days in ascending order according to the fraction of visitors who visited the waterfall.

   Ⓐ Tuesday, Wednesday, Monday
   Ⓑ Monday, Wednesday, Tuesday
   Ⓒ Wednesday, Tuesday, Monday
   Ⓓ Monday, Tuesday, Wednesday

15. What makes 4/9 > 4/11?

   Ⓐ The numerators are the same, so the first fraction is automatically greater.
   Ⓑ Ninths are smaller than elevenths.
   Ⓒ Ninths are larger than elevenths.
   Ⓓ Ninths and elevenths are the same size.

Name _____   Date _____

Section 2: Build Fractions from unit fractions

Adding and Subtracting Fractions (4.NF.B.3.A)

1. Find the sum:
   3/12 + 2/12 =

   Ⓐ 5/24
   Ⓑ 6/12
   Ⓒ 5/12
   Ⓓ 1/4

2. Find the sum:
   2/7 + 1/7 =

   Ⓐ 3/7
   Ⓑ 3/14
   Ⓒ 2/14
   Ⓓ 1/3

3. Use the following model to determine what fractional parts are shaded.

   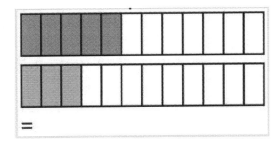

   Ⓐ 1/2
   Ⓑ 5/12
   Ⓒ 2/3
   Ⓓ 3/4

Name _____ Date _____

4. What fraction expresses the total amount represented by the fraction models?

Ⓐ 3/5
Ⓑ 3/10
Ⓒ 1/2
Ⓓ 3/10

5. Mary cut a pizza into 9 pieces. Mary ate 3 pieces, and Connie ate 3 pieces. What part of the pizza did they eat all together?

Ⓐ 9/18
Ⓑ 2/3
Ⓒ 6/18
Ⓓ 3/4

Name _____ Date _____

Adding and Subtracting Fractions through Decompositions (4.NF.B.3.B)

1. 1 1/4 - 3/4 = ☐

   Ⓐ 6/4 – 3/4 = 3/4
   Ⓑ 4/4 – 3/4 = 1/4
   Ⓒ 4/8 – 3/4 = 1/4
   Ⓓ 5/4 – 3/4 = 2/4 (1/2)

2. How many sevenths are there in 3 whole pizza?

   Ⓐ 7
   Ⓑ 14
   Ⓒ 28
   Ⓓ 21

3. What fractional part could be added to each blank to make each number sentence true?

   3/8 = 1/8 + ____ + ____;
   3/8 = ____ + 2/8

   Ⓐ 2/8
   Ⓑ 1/8
   Ⓒ 0/8
   Ⓓ 3/8

4. 2 3/5 – 4/5 =

   Ⓐ 1 1/10
   Ⓑ 1 4/5
   Ⓒ 1 2/5
   Ⓓ 1 1/5

5. How many sixths are there in 6 birthday cakes?

   Ⓐ 30
   Ⓑ 40
   Ⓒ 36
   Ⓓ 25

Adding and Subtracting Mixed Numbers (4.NF.B.3.C)

1. Angelo picked 2 3/4 pounds of apples from the apple orchard. He gave 1 1/4 pounds to his neighbor Mrs. Mason. How many pounds of apples does Angelo have left?

    Ⓐ 1 1/2 pounds
    Ⓑ 1 3/4 pounds
    Ⓒ 2 1/4 pounds
    Ⓓ 1 3/8 pounds

2. Daniel and Colby are building a castle out of plastic building blocks. They will need 2 1/2 buckets of blocks for the castle. Daniel used to have two full buckets of blocks, but lost some, and now only has 1 3/4 buckets. Colby used to have two full buckets of blocks too, but now has 1 1/4 buckets. If Daniel and Colby combine their buckets of blocks, will they have enough to build their castle?

    Ⓐ No, they will have less than 1 1/2 buckets.
    Ⓑ No, they will have 1 1/2 buckets.
    Ⓒ Yes, they will have 2 1/2 buckets.
    Ⓓ Yes, they will have 3 buckets.

3. Lexi and Ava are making chocolate chip cookies for a sleepover with their friends. They will need 4 1/4 cups of chocolate chips to make enough cookies for their friends. Lexi has 2 3/4 cups of chocolate chips. Ava has 1 3/4 cups of chocolate chips. Will the girls have enough chocolate chips to make the cookies for their friends?

    Ⓐ They'll have less than 4 cups, but should just use the amount they have.
    Ⓑ They'll have less than 4 cups, so no.
    Ⓒ They'll have 4 1/4 cups, so yes.
    Ⓓ They'll have 4 2/4 cups, so yes.

4. 3 2/4 + 1 1/4 =

    Ⓐ 20/4
    Ⓑ 2 3/4
    Ⓒ 4 3/4
    Ⓓ 5 3/4

Name _____ Date _____

5. 7 9/9 − 3 5/9 =

- Ⓐ 2 7/9
- Ⓑ 3 4/9
- Ⓒ 3 3/9
- Ⓓ 4 4/9

Name _____   Date _____

Adding and Subtracting Fractions in Word Problems (4.NF.B.3.D)

1. Marcie and Lisa wanted to share a cheese pizza together. Marcie ate 3/6 of the pizza, and Lisa ate 2/6 of the pizza. How much of the pizza did the girls eat together?

    Ⓐ 6/6 of a pizza
    Ⓑ 5/6 of a pizza
    Ⓒ 1/2 of a pizza
    Ⓓ 4/6 of a pizza

2. Sophie and Angie need 8 5/8 feet of ribbon to package gift baskets. Sophie has 3 1/8 feet of ribbon and Angie has 5 3/8 feet of ribbon. Will the girls have enough ribbon to complete the gift baskets?

    Ⓐ Yes, and they will have extra ribbon.
    Ⓑ Yes, but they will not have extra ribbon.
    Ⓒ They will have just enough ribbon to make the baskets.
    Ⓓ No, they will not have enough ribbon to make the baskets.

3. Travis has 4 1/8 pizzas left over from his soccer party. After giving some pizza to his friend, he has 2 4/8 of a pizza left. How much pizza did Travis give to his friend?

    Ⓐ 1 1/2 pizzas
    Ⓑ 1 5/8 pizzas
    Ⓒ 1 3/4 pizzas
    Ⓓ 1 5/7 pizzas

4. Which student solved the problem correctly?

Student 1	Student 2	Student 3
3 + 2 = 5 and 3/4 + 1/4 = 1 so 5 + 1 = 6	3 3/4 + 2 = 5 3/4 + 1/4 = 6	3 3/4 = 1 5/4 and 2 1/4 = 9/4 so 15/4 + 9/4 = 24/4 = 6

Ⓐ Student 1
Ⓑ Student 2
Ⓒ Student 3
Ⓓ All of the students

5. There are 2 loaves of freshly baked bread, and each loaf is cut into 8 equal pieces. If 5/8 of a loaf is used for breakfast, and 7/8 of a loaf is used for lunch, what fraction of the bread if left?

Ⓐ 1/4 of a loaf
Ⓑ 2/16 of a loaf
Ⓒ 1/8 of a loaf
Ⓓ 1/10 of a loaf

Name _____   Date _____

Multiplying Fractions (N.F.B.4.A)

1. Solve 1/2 x 6 =

    A) 2/6
    B) 1/3
    C) 3
    D) 3/6

2. Solve 6 x 1/4 =

    A) 2
    B) 1 1/2
    C) 4/6
    D) 3

3. What product do these models show?

    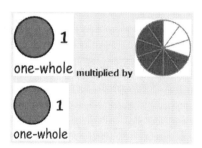

    A) 4/10
    B) 14/100
    C) 7/10
    D) 1 4/10

4. Find the product:
   45 x 2/3

    A) 90
    B) 86
    C) 30
    D) 129

5. When she took her children shopping, Mrs. Shontz noticed that all 4 of their feet had grown 1/2 inch. What is the total number of inches the Shontz children's feet grew?

Ⓐ 1/2 inch
Ⓑ inch
Ⓒ 1 1/2 inches
Ⓓ 2 inches

Multiplying Fractions by a Whole Number (4.NF.B.4.B)

1. 7 x 1/3 =

    Ⓐ 1/21
    Ⓑ 3/7
    Ⓒ 7/22
    Ⓓ 7/3

2. Cook is making sandwiches for the party. If each person at a party eats 2/8 of a pound of turkey, and there are 5 people at the party, how many pounds of turkey are needed? Between what two whole numbers does your answer lie?

    Ⓐ 3 pounds
    Ⓑ 1 1/4 pounds
    Ⓒ 2 1/4 pounds
    Ⓓ 4/8 pounds

3. 1/5 x 7 =

    Ⓐ 7/5
    Ⓑ 5/7
    Ⓒ 35
    Ⓓ 1/7

4. 1/9 x 8 =

    Ⓐ 72
    Ⓑ 8/9
    Ⓒ 1/8
    Ⓓ 9/8

5. Which product is greater than 1?

    Ⓐ 2 x 2/3
    Ⓑ 4 x 1/6
    Ⓒ 3 x 2/10
    Ⓓ 2 x 2/5

Multiplying Fractions in Word Problems (4.NF.B.4.C)

1. Aimee is making treat bags for her Christmas party. She is going to put 2/3 cups of mint M&Ms in each bag. She has invited 9 friends to her party. How many cups of mint M&Ms does she need for her friends' treat bags?

   Ⓐ 5 cups
   Ⓑ 6 cups
   Ⓒ 4 cups
   Ⓓ 5 1/2 cups

2. Aimee was making treat bags from Question #1, but then she decided to also include 1/2 of a cup of coconut M&Ms in each bag. How many cups of coconut M&Ms does she need? (Remember that she has 9 friends.)

   Ⓐ 4 1/2 cups
   Ⓑ 4 cups
   Ⓒ 3 1/3 cups
   Ⓓ 3 cups

3. Aimee's mom bought palm tree bags to celebrate their move to Florida. This is their first Christmas in Florida. Each treat bag will hold one cup of treats. Aimee wants to use 2/3 cup mint M&M's and 1/2 cup coconut M&M's. Will Aimee be able to fit all of the M&Ms in each party bag for her friends?

   Ⓐ Yes
   Ⓑ No

4. Kendra runs 3/4 mile each day. How many miles does she run in 1 week?

   Ⓐ 5 miles
   Ⓑ 5 1/4 miles
   Ⓒ 5 1/2 miles
   Ⓓ 5 3/4 miles

5. Mrs. Howett is making punch. The punch uses 3/5 cup of grapefruit juice for one serving. If she makes 4 servings, how many cups of grapefruit juice does she need?

   Ⓐ 2 2/5 cups
   Ⓑ 1 2/5 cups
   Ⓒ 3 1/5 cups
   Ⓓ 1 4/5 cups

Name _____   Date _____

Section 3: Understand decimal notation for fractions, and compare decimal fractions

10 to 100 Equivalent Fractions (4.NF.C.5)

1. What fraction of a dollar is 6 dimes and 3 pennies?

   Ⓐ 0.63 or 63/100
   Ⓑ 0.73 or 73/100
   Ⓒ 0.53 or 53/100
   Ⓓ 0.43 or 43/100

2. 1 tenth + 4 hundredths = _____ hundredths

   Ⓐ 14
   Ⓑ 140
   Ⓒ 1400
   Ⓓ 104

3. 4 hundredths + 1 tenth = _____ hundredths

   Ⓐ 140
   Ⓑ 14
   Ⓒ 141
   Ⓓ 41

4. 5 tenths + 2 hundredths = _____ hundredths

   Ⓐ 25
   Ⓑ 525
   Ⓒ 52
   Ⓓ 502

5. 5 hundredths + 2 tenths = _____ hundredths

   Ⓐ 25
   Ⓑ 52
   Ⓒ 252
   Ⓓ 502

Name _____   Date _____

6. 14 hundredths = _____ hundredths + 4 hundredths

   Ⓐ 144
   Ⓑ 414
   Ⓒ 104
   Ⓓ 10

7. 14 hundredths = _____ tenths + 4 hundredths

   Ⓐ 10
   Ⓑ 100
   Ⓒ 1
   Ⓓ 0

8. 14 hundredths = 1 tenth + 3 hundredths + _____ hundredths

   Ⓐ 10
   Ⓑ 1
   Ⓒ 0
   Ⓓ 4

9. 80 hundredths = _____ tenths

   Ⓐ 8
   Ⓑ 80
   Ⓒ 0
   Ⓓ 1

10. 2/10 + 41/100 =

   Ⓐ 43/100
   Ⓑ 43/10
   Ⓒ 61/100
   Ⓓ 61/10

Convert Fractions to Decimals (4.NF.C.6)

1. Point A is located closest to _____ on this number line?

   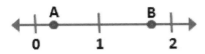

   Ⓐ 0
   Ⓑ 0.75
   Ⓒ 0.25
   Ⓓ -1

2. Convert 148/1,000 to a decimal.

   Ⓐ 0.148
   Ⓑ 14.8
   Ⓒ .00148
   Ⓓ 0.0148

3. Where is Point D located on this number line?

   Ⓐ -2.5
   Ⓑ -2
   Ⓒ -1.5
   Ⓓ -3

4. Convert 129 1/4 to decimal.

   Ⓐ 129.25
   Ⓑ 129.025
   Ⓒ 129.75
   Ⓓ 129.14

5. Convert the fraction to a decimal: 3/1,000 = _____

   Ⓐ 0.03
   Ⓑ 0.003
   Ⓒ 0.30
   Ⓓ 0.0003

6. What decimal does this model represent?

   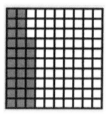

   Ⓐ 2.7
   Ⓑ 2.0
   Ⓒ 0.207
   Ⓓ 0.27

7. Mr. Hughes preferred to convert this decimal into a fraction in order to add it to a group of other fractions. Which answer is correct?
   0.044

   Ⓐ 44/100
   Ⓑ 44/10,000
   Ⓒ 44/1,000
   Ⓓ 44/10

8. Convert the mixed number to a decimal: 1 4/1,000 = _____

   Ⓐ 1.04
   Ⓑ 1.004
   Ⓒ 1.4
   Ⓓ .1004

9. The construction crew was working 0.193 of the times that we traveled that road. What fraction of the time was the crew working?

   Ⓐ 193/100
   Ⓑ 193/10,000
   Ⓒ 193/1,000
   Ⓓ 93/100

10. What are the addends in this problem?
    0.300 + 0.249

    Ⓐ 300/100 + 249/100
    Ⓑ 300/100 + 249/1,000
    Ⓒ 549/1,000
    Ⓓ 300/1,000 + 249/1,000

Name _____ Date _____

Compare Decimals (4.NF.C.7)

1. Point B is located closest to ____ on this number line.

   Ⓐ 1.75
   Ⓑ 1.5
   Ⓒ 2.25
   Ⓓ 1.1

2. Compare the following decimals using <, >, or =.
   0.05 ___ 0.50

   Ⓐ 0.05 < 0.50
   Ⓑ 0.05 = 0.50
   Ⓒ 0.05 > 0.50

3. The rainbow trout dinner costs $23.99. The steak dinner costs $26.49. The roast chicken dinner is $5.00 less than $30.79. Which dinner costs the most?

   Ⓐ rainbow trout dinner
   Ⓑ steak dinner
   Ⓒ chicken dinner

4. Compare the following decimals using <, >, or =.
   0.2 ___ 0.200

   Ⓐ 0.2 < 0.200
   Ⓑ 0.2 = 0.200
   Ⓒ 0.2 > 0.200

5. Which of these decimals is the greatest?
   0.0060
   0.006
   0.060

   Ⓐ 0.0060
   Ⓑ 0.006
   Ⓒ 0.060

6. Which comparison symbol makes this statement true?
   1.954 ___ 0.1954

   Ⓐ 1.954 = 0.1954
   Ⓑ 1.954 > 0.1954
   Ⓒ 1.954 < 0.1954

7. Order these decimals from least to greatest.
   43.75; 0.4385; 0.04375

   Ⓐ 0.04375; 43.75; 0.4385
   Ⓑ 0.4385; 0.04375; 43.75
   Ⓒ 0.04375; 0.4385; 43.75
   Ⓓ 0.4385; 43.75; 0.04375

8. Compare the following decimals using <, >, or =.
   1.10 ___ 1.1000

   Ⓐ 1.10 < 1.1000
   Ⓑ 1.10 = 1.1000
   Ⓒ 1.10 > 1.1000

9. A one-way airline ticket to Atlanta from New York costs $65.00 more on the weekend than it does during the week. How much would a $225 (weekday price) ticket cost if the traveler needed to fly on a Saturday?

   Ⓐ $270.00
   Ⓑ $290.00
   Ⓒ $280.00
   Ⓓ $300.00

Name _____ Date _____

10. A pattern exists in the prices of the following vehicles. Which numbers complete this table?

Vehicle	Price
compact car	$30,000.00
midsize car	$35,000.00
luxury car	$40,000.00
jeep	?
mini van	$50,000.00
full size van	?

Ⓐ $36,000.00; $51,000.00
Ⓑ $45,000.00; $55,000.00
Ⓒ $40,000.00; $51,000.00
Ⓓ $36,000.00; $55,000.00

End of Number & Operations - Fractions

Number & Operations - Fractions

Answer Key
&
Detailed Explanations

Equivalent Fractions (4.NF.A.1)

Question No.	Answer	Detailed Explanation
1	C	The denominator (bottom number) is the total number of items presented. The numerator (top number) is the number of identified items. The phrase "what fraction" also means "how many".
2	C	There are three different shapes represented. This question is asking for the number of square and triangles. That number of shapes that are not circles is the numerator and the total number of shapes is the denominator.
3	A	The fraction should only pertain to the number of squares: the number of shaded squares is the numerator and the total number of squares is the denominator.
4	D	The number of shaded circles is the numerator and the total number of **shaded** shapes is the denominator.
5	C	All of these fractions represent 1/2. The numerators are 1 part out of 2 parts: 4 is two parts of 2. 6 is two parts of 3.
6	B	The correct answer would be fractions which have numerators with a sum of 11 and denominators that are both 12.
7	D	The model represents 1 part of something that is divided into 4 equal pieces. An equivalent fraction would also be 1/4 of a total number of parts.
8	D	Draw a model of 8/18. Choose the fraction that has the same portion sizes as 8/18.
9	B	Each equivalent fraction represents 5 parts out of 6, and when a GCF is found, it will be multiplied by 5 and the product will be the numerator of the fraction and multiplied by 6 and the product will be the denominator of the fraction. For example: 2 x 5 = 10 and 2 x 6 = 12.
10	B	Find the GCF. This is the largest number that both the numerator and denominator can be divided by and will give the numerator and denominator as products. The quotients are the numerator and denominator reduced to its lowest terms: for example, 15/20 is reduced to 3/4 because 15 is divided by 5 (GCF) 3 times and 20, 4 times. Five is the largest number that 15 and 20 can be divided by evenly.
11	D	Find the GCF, which is the largest factor that both the numerator and denominator can be divided by.

Question No.	Answer	Detailed Explanation
12	B	The correct fraction can be reduced to its lowest terms of 4/5: Find the Greatest Common Factor (GCF), which is a number that the numerator and denominator can be divided by: 80 divided by **20** = 4 and 100 divided by **20** = 5. In this case, the GCF is 20. The number of times the numerator and denominator divides evenly into the GCF (4/5) is the lowest terms. 60/75 also reduces to 4/5 when reduced to lowest terms. (GCF = 15)
13	B	Reduce the fraction to its lowest terms by dividing the numerator and denominator by the GCF (9).
14	C	Use the GCF of the numerator and denominator of each fraction to determine if it is equivalent to 2/9.
15	B	These fractions all reduce to 5/8 in their lowest terms.

Compare Fractions (4.NF.A.2)

1	C	Point D is located halfway between -1 and -2 on the number line. This point would represent -1.5, or -1 1/2.
2	A	All of these fractions are equivalent because they reduce to the same fraction when reduced to lowest terms. They would all reduce to 1/7.
3	C	Both fractions are showing three parts. 3/12 is 3 parts out of 12; 3/18 is 3 parts out of 18. The greater the denominator, the more pieces the wholes have been divided into. Therefore, the pieces will be smaller.
4	B	Four pieces from a cake that has been divided into 20 equal pieces will be more cake than 4 pieces cut from a cake divided into 28 equal pieces.
5	A	Add the numerators to compare the actual sum with the third fraction.
6	A	Subtract the numerators to compare the numerator with the third fraction.
7	C	The model above is divided into 12 equal parts. If written as a fraction, the denominator would be 12. The greater the denominator, the more pieces the pie or whole is divided into, and therefore, the smaller the pieces. 2/10 would be greater than 2/12.
8	B	The fractions should get smaller. Compare the number of parts (numerator) of a whole (denominator).

Question No.	Answer	Detailed Explanation
9	B	The shaded area in Model D is the greatest. The shaded area in Model B is the least.
10	A	Add the numerators and compare the sums.
11	A	The denominators of these fractions are all the same, so the numerators should be listed from least to greatest. 13 could go between 9 and 16.
12	C	Subtract 3/7 from 6/7. 6/7 - 3/7 = 3/7. Dan eats vegetables 3/7 of a week (or 3 days) less than Marty.
13	C	1/2 is the only fraction listed that is greater than 1/3.
14	B	The denominators are all the same, so use the numerators of the fractions to list them in increasing order.
15	C	If a whole is divided into 9 equal pieces and another congruent whole is divided into 11 equal pieces, the elevenths will be smaller than the ninths.

Adding and Subtracting Fractions (4.NF.B.3.A)

1	C	When denominators are the same, showing a portion that is divided into the same number of parts, add the numerators. The numerator is the sum of the numerators, and the denominator remains the same.
2	A	The denominators are the same, so add the numerators.
3	C	The shaded sections represent the numerator. The number of sections the models are divided up into (counted one time) is the denominator. In first row, there are 5/12 shaded, and in the second row, there are 3/12 shaded which equals 8/12. 8 can be divided by 4 which equals 2. 12 can be divided by 4 which equals 3. Therefore, 8/12 can be reduced to 2/3. Reduce the fraction to its lowest terms.
4	A	Add up all of the shaded areas; this is the numerator. 3/10 plus 3/10 would need to be added together. Since the denominator is the same, only the numerator needs to be added (6). So, the answer is 6/10; 6/10 can be divided by 2, for 3/5. The denominator is the number of sections a model is divided into. Reduce that fraction to its lowest terms.
5	B	The number of pieces the pizza is cut into is 9, which is the denominator. Each girl ate 3 pieces. The total number of pieces eaten is the numerator, which is 6. 6/9 is the answer. 6/9 can be divided evenly by 3, so the answer is 2/3. Reduce the fraction to its lowest terms.

Name _____ Date _____

Adding and Subtracting Fractions through Decompositions (4.NF.B.3.B)

Question No.	Answer	Detailed Explanation
1	D	One whole would be 4/4, and you then add an additional 1/4, for a total of 5/4. Then, subtract 3/4 from 5/4. Reduce to 1/2.
2	D	One whole would be 7/7. Two wholes would be 7/7 + 7/7= 14/7. Three whole would be 21/7.
3	B	Each number sentence should equal 3/8. In the first sentence, 1/8 needs to be added twice, and in the second sentence once.
4	B	Explanation: 2 3/5 is equal to 13/5. 13/5- 4/5 equals 9/5, or, 1 4/5.
5	C	6 cakes multiplied by 6 pieces in each cake would be 36 pieces

Adding and Subtracting Mixed Numbers (4.NF.B.3.C)

1	A	To determine how many pounds of apples he has left, subtract the amount of apples he gave away from the amount of apples he picked. Break this into two smaller subtraction problems, beginning with the whole numbers. Angelo had 2 full pounds of apples, and gave 1 full pound away: 2−1=1. He now has 1 full pound of apples. In addition to the 2 full pounds, he had 3/4 of a pound of apples, and gave 1/4 of a pound away: 3/4−1/4=2/4. He now has 2/4 of a pound of apples in addition to the 1 full pound of apples. All together, Angelo has 1 2/4 pounds of apples left. This can be reduced to 1 1/2 pounds.
2	D	1 3/4+1 1/4=1+3/4+1+1/4. Using the Commutative Property of Addition, we have that 1+3/4+1+1/4=1+1+3/4+1/4. We can then simply, & add the whole numbers together, & add the fractions together. Adding the whole numbers we have 1+1=2. Adding the fractions we have 3/4+1/4=4/4=1. Together we have 2+1=3. Combined, the boys have 3 buckets of plastic building blocks.
3	D	The amount of chocolate chips needed can be expressed in number of fractional pieces (18/4), or number of wholes, and additional fractional pieces (4 2/4). This task requires an understanding of comparing fractions or mixed numbers. Combine 2 3/4 and 1 3/4. 1/4 plus 3/4 can be combined to make a new whole. The combined total of chips will be 4 2/4 cups. Because 4 2/4 is greater than 4 1/4, Lexi and Ava will have enough chocolate chips for their cookies. Students may try to (incorrectly) reduce 4 2/4 to 4 1/4, it will reduce to 4 1/2 though. These measurements may be difficult for students to visualize.

Question No.	Answer	Detailed Explanation
4	C	Add the whole numbers (3) and (1) to get (4). Then add 2/4 and 1/4 to equal 3/4.
5	D	Subtract whole numbers (7) – (3) for an answer of 4. Then subtract numerators Since the demoninators are the same: (9)- (5) for an answer of 3. Combine answers: 4 4/9.

Adding and Subtracting Fractions in Word Problems (4.NF.B.3.D)

1	B	Explanation: The amount of pizza Marcie ate can be thought of a 3/6 (or 1/6 and 1/6 and 1/6). The amount of pizza Lisa ate can be thought of a 1/6 and 1/6. The total amount of pizza they ate is 1/6 + 1/6 + 1/6 + 1/6 + 1/6 or 5/6 of the whole pizza.
2	D	Explanation: The ribbon Sophie has can be added to the ribbon Angie has to determine how much ribbon they have altogether. Sophie has 3 1/8 feet of ribbon and Avery has 5 3/8 feet of ribbon. This can be written as 3 1/8 + 5 3/8. The girls have 8 feet of ribbon: (You add the 3 and 5, and they also have 1/8 and 3/8 which makes a total of 4/8 more). Altogether the girls have 8 4/8 feet of ribbon, 8 4/8 is less than 8 5/8 so they will not have enough ribbon to complete the project. They will be short 1/8 of a foot of ribbon.
3	B	Explanation: Travis had 4 1/8 pizzas to start. This is 33/8 of a pizza. The xs show the pizza he has left which is 2 4/8 pizzas or 20/8 pizzas. The shaded rectangles without the xs are the pizza he gave to his friend which is 13/8 or 1 5/8 pizzas.
4	D	Explanation: All ways are correct ways to solve the problem; any answer given above is correct.
5	A	Each loaf contains 8 pieces, so there are 16 pieces total. 5/8 + 7/8 have been used so far, which totals 12/16. 16 pieces minus 12 pieces eaten equals 4 pieces left (out of 16), so 4/16 reduces to 1/4.

Name _____   Date _____

Multiplying Fractions (4.NF.B.4.A)

Question No.	Answer	Detailed Explanation
1	C	Multiply 1/2 x 6/1. The answer is a fraction in which the numerator is larger than the denominator. This is called an improper fraction, which is always reduced to its lowest terms by dividing the denominator into the numerator: 6/2. The quotient will be a whole number if there is no remainder. This is the answer.
2	B	The first answer is an improper fraction that, when divided into the numerator, has a remainder. Therefore, the answer will be a mixed fraction-a whole number and a fraction. The whole number represents how many times the denominator divides into the numerator. The remainder represents the numerator in the fraction and the denominator is the same denominator of the improper fraction in the original problem.
3	D	Convert the whole number into a fraction: 2/1. Multiply the numerators together and the denominators together. The answer is an improper fraction. To reduce it to its lowest terms, divide the denominator into the numerator. If there is no remainder, the answer is a whole number. If there is a remainder, the answer will be a mixed number. The remainder is the fraction's numerator, and the denominator is the same as the one for the improper fraction.
4	C	Convert the whole number into a fraction: 45/1. Multiply the numerators together, then the denominators together. Reduce the answer to the lowest terms.
5	D	Multiply 4 x 1/2 by converting the 4 into a fraction: 4/1. Reduce the answer to its lowest terms.

Multiplying Fractions by a Whole Number (4.NF.B.4.B)

1	D	To multiply a whole number by a unit fraction, multiply the whole number by 1 to get the numerator in the product. The denominator will not change.
2	B	Add up all of the fractional parts of meat. 2/8 + 2/8 + 2/8 + 2/8 + 2/8 = 10/8 pounds. Then convert the improper fraction to a mixed number: 1 2/8 or 1 1/4 pounds.

Name _____   Date _____

Question No.	Answer	Detailed Explanation
3	A	To multiply a whole number by a unit fraction, multiply the whole number by 1 to get the numerator in the product. The denominator will not change.
4	B	To multiply a whole number by a unit fraction, multiply the whole number by 1 to get the numerator in the product. The denominator will not change.
5	A	All answers except for (A) will equal less than 1. 2 x 2/3 equals 1 1/3.

Multiplying Fractions in Word Problems (4.NF.B.4.C)

1	B	Students may use the multiplication equation 9 X 2/3 =18/3 or 6 cups of mint M&Ms. Students may use the repeated addition equation 2/3 +2/3+2/3+2/3+2/3+2/3+2/3+2/3+2/3=18/3 or 6 cups of mint M&Ms.
2	A	Students could have used: 9 X 1/2= 9/2 or 4 1/2 cups of coconut M&Ms. Or, students could have used: 1/2 + 1/2 + 1/2 +1/2 +1/2 + 1/2 + 1/2 + 1/2 + 1/2 =9/2 or 4 1/2 cups of coconut M&Ms.
3	B	Student may draw a model that shows that 2/3 + 1/2 >1. Student may state that 2/3 is more than 1/2. If 1/2 + 1/2 = 1, then 2/3 + 1/2 > 1. Student may convert fractions to common denominators and add to find the total is 7/6 or 1 1/6 cups of M&Ms.
4	B	3/4 x 7= 5 1/4 miles. 3/4+ 3/4+3/4+ 3/4+ 3/4 +3/4+3/4= 5 1/4 miles.
5	A	3/5 x 4= 2 2/5, or 3/5 + 3/5+ 3/5 + 3/5 + 3/5= 2 2/5

10 to 100 Equivalent Fractions (4.NF.C.5)

1	A	6 dimes is 60 pennies, so 6 dimes and 3 pennies is 63 pennies, which is 63/100=0.63 of a dollar
2	A	thousands | hundreds | tens | units | . decimal | tenths | hundredths | thousandths | ten thousandths 1 | 4

Name _____ Date _____

Question No.	Answer	Detailed Explanation
3	B	thousands | hundreds | tens | units | . decimal | tenths | hundredths | thousandths | ten thousandths tenths: 1, hundredths: 4
4	C	tenths: 5, hundredths: 2
5	A	tenths: 2, hundredths: 5
6	D	tenths: 10
7	C	tenths: 1

Name _____ Date _____

Question No.	Answer	Detailed Explanation
8	B	(place value chart: thousands, hundreds, tens, units, decimal, tenths, hundredths, thousandths, ten thousandths — with 1 in the ten thousandths column)
9	A	(place value chart: thousands, hundreds, tens, units, decimal, tenths, hundredths, thousandths, ten thousandths — with 8 in the tenths column)
10	C	Students must realize that 2/10 is equivalent to 20/100 and then add add 20/100+41/100.

Convert Fractions to Decimals (4.NF.C.6)

1	C	Point A is located near 1/4 of 1 on the number line. In order to represent numbers that are less than 1, a fraction or decimal is used. To convert the fraction to a decimal, divide the numerator by the denominator, adding zeroes until there is no remainder. A decimal point is plotted after the last digit in the dividend. Another decimal point is plotted directly above that one in the answer. The decimal equivalent of 1/4 is 0.25.
2	A	This fraction would be read "one hundred forty-eight thousandths." As a decimal, that number is written 0.148.
3	C	Point D is halfway between -1 and -2. Halfway between -1 and -2 is -1.5.
4	A	Convert the fraction part of this problem into a decimal by dividing the denominator into the numerator: 1 divided by 4. Put the whole number part of this problem in front of the decimal to complete the answer.
5	B	This fraction would be read "three thousandths." As a decimal, it is written 0.003.
6	D	The fraction this model represents is 27/100: there are 27 cells shaded and 100 cells altogether. Twenty-seven hundredths is written as 0.27 in decimal form.

84

© Lumos Information Services 2015 | LumosLearning.com

Name _____ Date _____

Question No.	Answer	Detailed Explanation
7	C	The last digit in this decimal is in the thousandths place. 44 thousands as a fraction would be written 44/1,000.
8	B	Divide the denominator into the numerator to change fraction to a decimal. Put the whole number in front of the decimal.
9	C	The last digit in the decimal is in the thousandths place. So, 0.193 would written as 193/1,000 in fraction form.
10	D	Convert the decimals into fractions to be added. Both of these decimals end in the thousandths place. So the addends are 300/1,000 and 249/1,000.

Compare Decimals (4.NF.C.7)

1	A	The point is plotted at about 1 3/4. Convert this mixed fraction to a decimal. 1 3/4 = 1.75
2	A	0.05 is 5 hundredths. 0.50 is 50 hundredths. 0.05 < 0.50
3	B	All of the dinner prices consist of decimals that have numbers in front of the decimal point. Those numbers are whole numbers (dollars). To compare, simply look at the dollar amounts. The cents do not matter as long as the dollar amounts are all different. Note: The cost of the chicken dinner is $25.79. (30.79 - 5.00 = 25.79)
4	B	Zeroes can be added after the last digit in a decimal number without changing its value. Therefore, 0.2 = 0.200.
5	C	All of these decimals have the same digit in the tenths place (a 0). Compare the digits in the hundredths place. The number that does not have a zero in the hundredths place has the greatest value.
6	B	Compare the digits in the ones place. The number with a non-zero digit in the ones place has the greater value. 1.954 > 0.1954
7	C	Numbers in the ten**ths** place are greater than numbers in the hundred**ths** place, numbers in the hundred**ths** place are greater than numbers in the thousand**ths** place, numbers in the thousand**ths** place are greater than numbers in the **ten** thousand**ths** place, and whole numbers are the greatest of them. Use the placement of the 4 to compare and order the numbers.
8	B	The zeroes after the last digit in each number do not change its value. 1.10 = 1.1000.
9	B	Add $65 to the weekday price. $225 + 65 = $290
10	B	There is a pattern here that is related to the size of the vehicle and an increase of $5,000.00. Each vehicle in the table costs $5,000 more than the vehicle listed above it.

Name _____   Date _____

Measurement & Data

Section 1: Solve problems involving measurement and conversion of measurement.

Units of Measurement (4.MD.A.1)

1. What customary unit should be used to measure the weight of the table shown in the picture below?

   Ⓐ pounds
   Ⓑ inches
   Ⓒ kilograms
   Ⓓ tons

2. Which of the following is an appropriate customary unit to measure the weight of a small bird?

   Ⓐ grams
   Ⓑ ounces
   Ⓒ pounds
   Ⓓ units

3. Complete the following statement:
   A horse might weigh _____.

   Ⓐ about 500 pounds
   Ⓑ about 12 pounds
   Ⓒ about a gallon
   Ⓓ about 200 ounces

86

© Lumos Information Services 2015 | LumosLearning.com

Name _____   Date _____

4. Choose the appropriate customary unit to measure the length of a road.

   Ⓐ Yard
   Ⓑ Meter
   Ⓒ Kilometer
   Ⓓ Mile

5. Choose the appropriate unit to measure the height of a tall tree.

   Ⓐ miles
   Ⓑ yards
   Ⓒ centimeters
   Ⓓ gallons

6. Which of the following is not a customary unit?

   Ⓐ Kilograms
   Ⓑ Yards
   Ⓒ Pounds
   Ⓓ Miles

7. Which of the following is a customary unit that can be used to measure the volume of a liquid?

   Ⓐ yards
   Ⓑ fluid ounces
   Ⓒ milliliters
   Ⓓ pounds

8. Which is more, 18 teaspoons or 2 fluid ounces?

   Ⓐ 2 fluid ounces
   Ⓑ They are equal.
   Ⓒ 18 teaspoons

9. A math textbook might weigh _____ .

   Ⓐ 125 ounces
   Ⓑ 25 ounces
   Ⓒ 200 ounces
   Ⓓ 2 ounces

10. To make the medicine seem to taste better, Mom told Bonita she had to take tablespoons instead of teaspoons. How many tablespoons of medicine should Bonita take if the dose is 3 teaspoons?

   Ⓐ tablespoons
   Ⓑ 1 tablespoon
   Ⓒ 6 tablespoons
   Ⓓ 1/2 tablespoon

Name _____ Date _____

Measurement Problems (4.MD.A.2)

1. Arthur wants to arrive at soccer practice at 5:30 PM. He knows it takes him 42 minutes to walk to practice from his house. Estimate the time Arthur should leave his house to go to practice?

   Ⓐ 5:00 PM
   Ⓑ 4:45 PM
   Ⓒ 4:30 PM
   Ⓓ 3:45 PM

2. A baseball game began at 7:05 PM and lasted for 2 hours and 38 minutes. At what time did the game end?

   Ⓐ 9:43 PM
   Ⓑ 10:33 PM
   Ⓒ 9:38 PM
   Ⓓ 9:33 PM

3. 4 feet and 5 inches is the same as:

   Ⓐ 48 inches
   Ⓑ 53 inches
   Ⓒ 41 inches
   Ⓓ 65 inches

4. Amir bought two cowboy hats for $47, a pair of cowboy boots for $150, and a leather belt for $32. The tax was $13.74. He gave the cashier $300. How much change does she owe him?

   Ⓐ $242.74
   Ⓑ $13.74
   Ⓒ $57.26
   Ⓓ $257.26

5. Harriet needed 1/2 cup of milk for the white sauce, but she could only find her tablespoon to measure with. How many tablespoons of milk will she need?

   Ⓐ 4 tablespoons
   Ⓑ 6 tablespoons
   Ⓒ 8 tablespoons
   Ⓓ 10 tablespoons

Name _____   Date _____

6. Use a comparison symbol to complete the following statement:
   32 ounces ___ 1 pound

   A. <
   B. =
   C. >

7. Rachel's gymnastics lessons lasted for 1 year. Sharon's lessons lasted for 9 months. Yolanda's lessons lasted for 23 months. How much time did the girls spend on lessons altogether?

   A. 4 years, 0 months
   B. 3 years, 0 months
   C. 3 years, 8 months
   D. 4 years, 8 months

8. To make some of the best cookies, mix 1 cup of butter, 2 cups of sugar, 2 1/2 cups of flour, 2 1/2 teaspoons of vanilla extract, 1/2 teaspoon of baking soda, and 3/4 cups of chocolate chips. Which comparison symbols would complete the following statements?
   amount of flour ___ amount of butter
   amount of butter ___ amount of sugar

   A. >; =
   B. <; >
   C. >; <
   D. <; =

9. The bookstore is selling paperback books for $3.25 each. How much would 4 paperbook books cost?

   A. $12.00
   B. $13.00
   C. $13.50
   D. $12.50

Name _____   Date _____

10. If Cindy bought 3 DVDs and 2 nacho kits, how much would she pay for all items before taxes? Use the table below to answer the question:

Item	Price
CDs	$10.99
DVDs	$24.99
Cordless Phone	$30.00
Flash Drives	$9.99
Nacho Kits	$6.99

Ⓐ $31.98
Ⓑ $74.97
Ⓒ $84.95
Ⓓ $88.95

Name _____ Date _____

Perimeter & Area (4.MD.B.3)

1. A rectangular room measures 10 feet long and 13 feet wide. How could you find out the area of this room?

    Ⓐ Add 10 and 13, then double the results
    Ⓑ Multiply 10 by 13
    Ⓒ Add 10 and 13
    Ⓓ None of the above

2. A rectangle has a perimeter of 30 inches. Which of the following could be the dimensions of the rectangle?

    Ⓐ 10 inches long and 5 inches wide
    Ⓑ 6 inches long and 5 inches wide
    Ⓒ 10 inches long and 3 inches wide
    Ⓓ 15 inches long and 15 inches wide

3. Which of these expressions could be used to find the perimeter of the above figure?

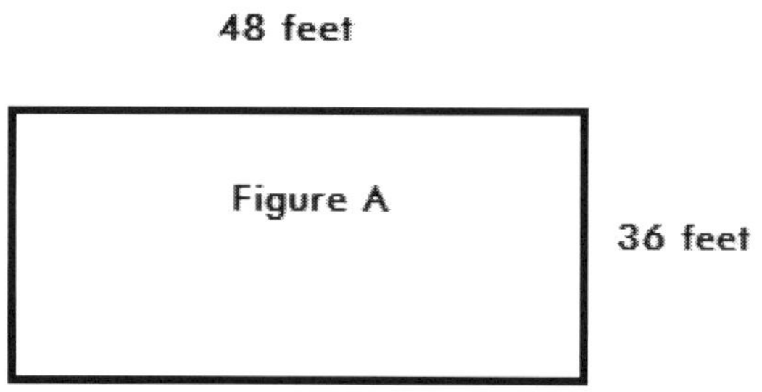

    48 feet

    Figure A   36 feet

    Ⓐ 48 + 36 + 2
    Ⓑ 48 x 36
    Ⓒ 2 x (48 + 36)
    Ⓓ 48 + 36

4. A chalkboard is 72 inches long and 30 inches wide. What is its perimeter?

    Ⓐ 204 inches
    Ⓑ 2,160 inches
    Ⓒ 102 inches
    Ⓓ 2,100 inches

Name _____ Date _____

5. Which of the following statements is true?

   Figure A   Figure B

   12 feet   10 feet
   [rectangle] 8 feet   [square] 10 feet

   Ⓐ The two shapes have the same perimeter.
   Ⓑ The two shapes have the same area.
   Ⓒ The two figures are congruent.
   Ⓓ Figure A has a greater area than Figure B.

6. If a square has a perimeter of 100 units, how long is each of its sides?

   Ⓐ 10 units
   Ⓑ 20 units
   Ⓒ 25 units
   Ⓓ Not enough information is given.

7. Find the perimeter of Shape C.

   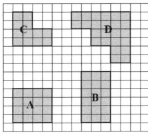

   ☐ = 1 square unit

   Ⓐ 8 units
   Ⓑ 12 units
   Ⓒ 14 units
   Ⓓ 16 units

8. Find the area of Shape C.

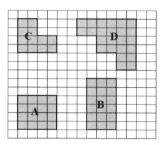

☐ = 1 square unit

Ⓐ 8 square units
Ⓑ 12 square units
Ⓒ 14 square units
Ⓓ 16 square units

9. What is the perimeter of this shape?

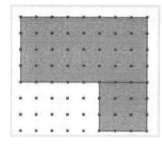

Ⓐ 24 units
Ⓑ 28 units
Ⓒ 30 units
Ⓓ 34 units

10. A rectangle has an area of 48 square units and a perimeter of 32 units. What are its dimensions?

Ⓐ 6 units by 8 units
Ⓑ 12 units by 4 units
Ⓒ 16 units by 3 units
Ⓓ All of the above are possible.

Section 2: Represent and interpret data

Representing and Interpreting Data (4.MD.B.4)

1. The students in Mrs. Riley's class were asked how many cousins they have. The results are shown in the line plot. Use the information shown in the line plot to respond to the following.
   How many of the students have no cousins?

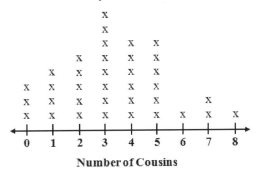

   Ⓐ 0 students
   Ⓑ 1 student
   Ⓒ 2 students
   Ⓓ 3 students

Name _____   Date _____

2. The students in Mrs. Riley's class were asked how many cousins they have. The results are shown in the line plot. Use the information shown in the line plot to respond to the following. How many of the students have exactly 4 cousins?

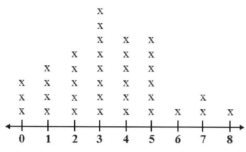

Ⓐ 1 student
Ⓑ 5 students
Ⓒ 6 students
Ⓓ 7 students

3. According to this graph, which foods are the 2 most favorite foods people enjoy at a carnival?
Favorite Carnival Foods

x			x			
x			x		x	x
x			x		x	x
x			x		x	x
x		x	x		x	x
x		x	x		x	x
x	x	x	x		x	x
x	x	x	x		x	x
x	x	x	x	x	x	x
x	x	x	x	x	x	x
x	x	x	x	x	x	x
caramel apples	elephant ears	corn dogs	cotton candy	french fries	candy apples	funnel cakes

Ⓐ candy apples and funnel cake
Ⓑ caramel apples and cotton candy
Ⓒ cotton candy and funnel cake
Ⓓ caramel apples and candy apples

4. How many more families prepare the night before the picnic than prepare right before leaving for the picnic?

x		
x	x	
x	x	x
x	x	x
Families That Pack the Night Before the Picnic	Families That Rise Early in the Morning to Pack for the Picnic	Families That Pack Right Before They Leave for the Picnic

(Each X is worth 2 points.)

Ⓐ 2
Ⓑ 4
Ⓒ 1
Ⓓ 8

5. The school nurse kept track of how many students visited her office during a week. She plotted the results on a line graph. Use the line graph to respond to the following. What trend should the nurse notice?

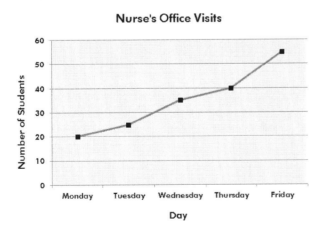

Ⓐ The number of students visiting her office increased throughout the week.
Ⓑ The number of students visiting her office decreased throughout the week.
Ⓒ The number of students visiting her office remained constant throughout the week.
Ⓓ There is no apparent trend.

Name _____ Date _____

6. The school nurse kept track of how many students visited her office during a week. She plotted the results on a line graph. Use the line graph to respond to the following. How many more students visited the nurse on Friday than on Monday?

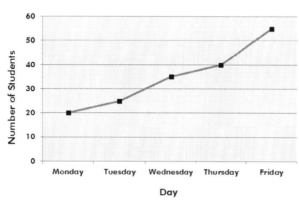

Ⓐ 20 students
Ⓑ 35 students
Ⓒ 45 students
Ⓓ 55 students

7. The school nurse kept track of how many students visited her office during a week. She plotted the results on a line graph. Use the line graph to respond to the following. How many students visited the nurse on Wednesday?

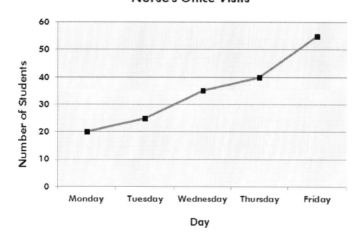

Ⓐ 30 students
Ⓑ 35 students
Ⓒ 40 students
Ⓓ 45 students

Name _____   Date _____

8. The fourth grade chorus is selling candy boxes to raise money for a trip to the water park. The pictograph below shows how many candy boxes they sold during the first four weeks of the sale. Use the information shown in the graph to respond to the following.
   The chorus needed to sell 150 boxes of candy to pay for the trip. Did they sell enough boxes?

   Candy Boxes Sold

Week 1	☐ ☐ ☐ ☐
Week 2	☐ ☐ ☐ ◺
Week 3	☐ ☐ ◺
Week 4	☐ ☐ ☐ ☐ ☐ ◺

   Key : ☐ = 10 boxes
         ◺ = 5 boxes

   Ⓐ Yes, they sold more than enough boxes.
   Ⓑ Yes, they sold exactly 150 boxes.
   Ⓒ No, they needed to sell 5 more boxes.
   Ⓓ No, they needed to sell 10 more boxes.

9. The fourth grade chorus is selling candy boxes to raise money for a trip to the water park. The pictograph below shows how many candy boxes they sold during the first four weeks of the sale. Use the information shown in the graph to respond to the following.
   How many more candy boxes were sold during the fourth week than during the first week?

   Candy Boxes Sold

Week 1	☐ ☐ ☐ ☐
Week 2	☐ ☐ ☐ ◺
Week 3	☐ ☐ ◺
Week 4	☐ ☐ ☐ ☐ ☐ ◺

   Key : ☐ = 10 boxes
         ◺ = 5 boxes

   Ⓐ 25 boxes
   Ⓑ 15 boxes
   Ⓒ 5 boxes
   Ⓓ None of the above

Name _____   Date _____

10. The fourth grade chorus is selling candy boxes to raise money for a trip to the water park. The pictograph below shows how many candy boxes they sold during the first four weeks of the sale. Use the information shown in the graph to respond to the following.
How many candy boxes were sold during the first two weeks of the sale?

Candy Boxes Sold

Week 1	☐ ☐ ☐ ☐
Week 2	☐ ☐ ☐ ◺
Week 3	☐ ☐ ◺
Week 4	☐ ☐ ☐ ☐ ☐ ◺

Key: ☐ = 10 boxes
◺ = 5 boxes

Ⓐ 1/2 boxes
Ⓑ 40 boxes
Ⓒ 35 boxes
Ⓓ 75 boxes

11. Mr. Green's class spent 5 weeks collecting cans as part of a recycling project. The bar graph shows how many cans they collected each week. Use the graph to respond to the following question.
How many cans were collected during the second week?

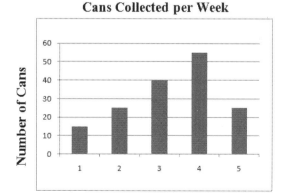

Ⓐ 15 cans
Ⓑ 20 cans
Ⓒ 25 cans
Ⓓ 35 cans

12. Mr. Green's class spent 5 weeks collecting cans as part of a recycling project. The bar graph shows how many cans they collected each week. Use the graph to respond to the following question.
How many more cans were collected during the third week than during the second week?

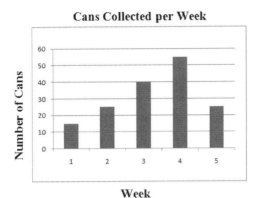

Ⓐ 5 cans
Ⓑ 15 cans
Ⓒ 25 cans
Ⓓ 40 cans

13. Mr. Green's class spent 5 weeks collecting cans as part of a recycling project. The bar graph shows how many cans they collected each week. Use the graph to respond to the following question.
How many cans were collected during the fourth week?

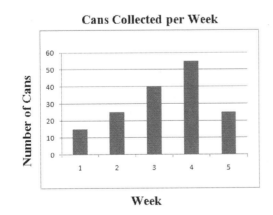

Ⓐ 65 cans
Ⓑ 55 cans
Ⓒ 50 cans
Ⓓ 40 cans

Name _____ Date _____

14. Mr. Green's class spent 5 weeks collecting cans as part of a recycling project. The bar graph shows how many cans they collected each week. Use the graph to respond to the following question.
During which week were the fewest cans collected?

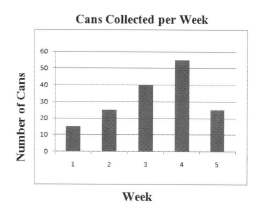

Ⓐ Week 1
Ⓑ Week 2
Ⓒ Week 3
Ⓓ Week 4

15. Mr. Green's class spent 5 weeks collecting cans as part of a recycling project. The bar graph shows how many cans they collected each week. Use the graph to respond to the following question.
During which two weeks were the same number of cans collected?

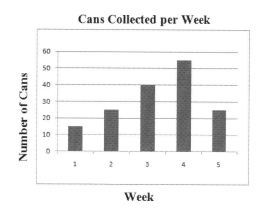

Ⓐ Weeks 1 and 2
Ⓑ Weeks 2 and 4
Ⓒ Weeks 3 and 5
Ⓓ Weeks 2 and 5

Name _____ Date _____

16. The students in the third grade were surveyed to find out their favorite seasons. The results are shown in the tally table. Use the tally table to respond to the following. Which season was chosen as the favorite of the most students?

Our Favorite Seasons

Winter	|||| |||| |||| ||
Spring	|||| ||
Summer	|||| |||| |||| |||
Fall	|||

- Ⓐ Winter
- Ⓑ Spring
- Ⓒ Summer
- Ⓓ Fall

17. The students in the third grade were surveyed to find out their favorite seasons. The results are shown in the tally table. Use the tally table to respond to the following. How many students chose winter as their favorite season?

Our Favorite Seasons

Winter	|||| |||| |||| ||
Spring	|||| ||
Summer	|||| |||| |||| |||
Fall	|||

- Ⓐ 12 students
- Ⓑ 16 students
- Ⓒ 17 students
- Ⓓ 22 students

18. The students in the third grade were surveyed to find out their favorite seasons. The results are shown in the tally table. Use the tally table to respond to the following. How many more students chose winter than spring?

Our Favorite Seasons

Winter	|||| |||| |||| ||
Spring	|||| ||
Summer	|||| |||| |||| |||
Fall	|||

- Ⓐ 5 students
- Ⓑ 10 students
- Ⓒ 15 students
- Ⓓ 17 students

19. The students in the third grade were surveyed to find out their favorite seasons. The results are shown in the tally table. Use the tally table to respond to the following. How many students were surveyed?

Our Favorite Seasons

Winter	|||| |||| |||| ||
Spring	|||| ||
Summer	|||| |||| |||| |||
Fall	|||

- Ⓐ 40 students
- Ⓑ 45 students
- Ⓒ 50 students
- Ⓓ 55 students

20. Teachers at a nearby elementary school took a survey to determine how to plan for future field trips. What is the sum of the field trip choices that received the highest votes?

Field Trips	No. of Votes
Jersey Cape	11
turtle Back Zoo	7
camden's Children Garden	8
NJ Adventure Aquarium	11

Ⓐ 19
Ⓑ 18
Ⓒ 22
Ⓓ 15

21. The chart shows the range of snowfall expected each month at a local ski resort. During which month is there the greatest range in the amount of snowfall?

Month	Inches of Snow Fall
November	0 - 5
December	5 - 15
January	10 - 50
February	10 - 20

Ⓐ November
Ⓑ December
Ⓒ January
Ⓓ February

22. Based on the bar graph below, what can be said about the trend in bicycle helmet safety?

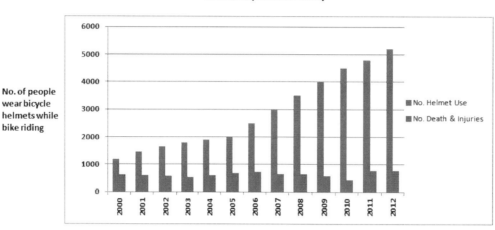

Trend in bicycle helmet safety

Ⓐ Bicycle riders who do not wear helmets are very unlikely to be injured or killed.
Ⓑ Bicycle riders who wear helmets are very unlikely to be injured or killed.
Ⓒ Wearing a helmet has no effect on bicycle safety.
Ⓓ Wearing a helmet effects bicycle safety on occassions.

23. The smoothie bar has several orders to fill. How many more orders call for sweet fruit smoothies than sour fruit smoothies?

Orders Call	Sweet Fruit Smoothies	Sour Fruit Smoothies
Ron Miller	10	2
Robert	5	5
George	4	3
Mike	6	4
Marisa	3	2
Greg	1	2

Ⓐ 23
Ⓑ 11
Ⓒ 12
Ⓓ 35

24. The crayon represented by a red x is longer than the crayon represented by a blue x by:

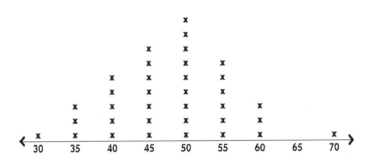

Ⓐ 1/2 inch
Ⓑ 1 inch
Ⓒ 1/4 inch
Ⓓ 3/4 inch

25. How many students studied for more than an hour?

Ⓐ 7
Ⓑ 10
Ⓒ 4
Ⓓ 1

Name _____   Date _____

26. The plot below shows the rainfall amounts (in inches) during the month of May. What is the difference the largest amount and the smallest amount of rain measured?

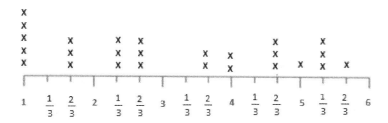

Ⓐ 4
Ⓑ 4 1/2
Ⓒ 4 3/4
Ⓓ 4 2/3

27. What is the length of this pencil?

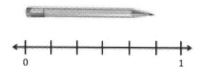

Ⓐ 1/2 of an inch
Ⓑ 5/6 of an inch
Ⓒ 4/6 of an inch
Ⓓ 2/6 of an inch

28. What is the length of this pencil?

Ⓐ 1/2 of an inch
Ⓑ 5/6 of an inch
Ⓒ 4/6 of an inch
Ⓓ 2/6 of an inch

29. What is the length of this pencil?

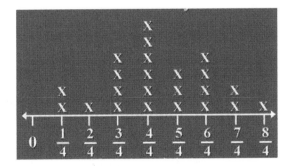

Ⓐ 1/2 of an inch
Ⓑ 5/6 of an inch
Ⓒ 4/6 of an inch
Ⓓ 2/6 of an inch

30. How many objects measured exactly 3/4"?

Ⓐ 5
Ⓑ 4
Ⓒ 3
Ⓓ 2

Name _____   Date _____

Section 3: Geometric measurement: understand concepts of angle and measure angles.

Angle Measurement (4.MD.C.5.A)

1. In the figure below, two lines intersect to form ∠ A, ∠ B, ∠ C, and ∠ D. If ∠ C measures 128°, then ∠ A measures:

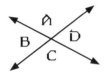

- Ⓐ 52°
- Ⓑ 128°
- Ⓒ 90°
- Ⓓ 180°

2. In the figure below, two lines intersect to form ∠ A, ∠ B, ∠ C, and ∠ D. If ∠ C measures 128°, then ∠ D measures:

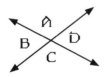

- Ⓐ 52°
- Ⓑ 128°
- Ⓒ 90°
- Ⓓ 180°

Name _____ Date _____

3. In the figure below, two lines intersect to form ∠ A, ∠ B, ∠ C, and ∠ D. If ∠ B measures 68°, then ∠ D measures:

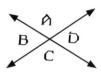

Ⓐ 152°
Ⓑ 90°
Ⓒ 180°
Ⓓ 68°

4. If the measurement of ∠ A is 64°, then what is the measurement of ∠ B?

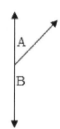

Ⓐ 126°
Ⓑ 119°
Ⓒ 116°
Ⓓ 64°

5. If the measurement of ∠ B is 94°, then what is the measurement of ∠ A?

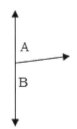

Ⓐ 90°
Ⓑ 86°
Ⓒ 66°
Ⓓ 94°

Name _____   Date _____

6. The total of the measures of ∠ A and ∠ B equals 90°. If ∠ A measures 51°, then what does ∠ B measure?

   Ⓐ 139°
   Ⓑ 51°
   Ⓒ 49°
   Ⓓ 39°

7. Two supplementary angles should total _____.

   Ⓐ 90°
   Ⓑ 45°
   Ⓒ 180°
   Ⓓ 360°

8. What should two complementary angles equal?

   Ⓐ 45°
   Ⓑ 90°
   Ⓒ 180°
   Ⓓ 360°

9. Complete the sentence:
   The measure of an obtuse angle is _____.

   Ⓐ less than the measure of a right angle
   Ⓑ equal to the measure of a right angle
   Ⓒ greater than the measure of a right angle
   Ⓓ less than the measure of an acute angle

10. What is the measure of the complement of an angle that measures 7°?

    Ⓐ 173°
    Ⓑ 87°
    Ⓒ 83°
    Ⓓ 73°

11. What is the measure of ∠A if ∠B measures 46°?

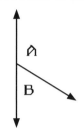

- Ⓐ 134°
- Ⓑ 143°
- Ⓒ 44°
- Ⓓ 90°

12. If ∠A measures 101° and ∠B measures 49° then what does ∠C measure?

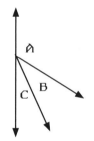

- Ⓐ 150°
- Ⓑ 20°
- Ⓒ 40°
- Ⓓ 30°

13. What is the supplement of an angle that measures 73°?

- Ⓐ 117°
- Ⓑ 107°
- Ⓒ 23°
- Ⓓ 17°

Name _____ Date _____

14. What is the measurement of each of the angles in this equilateral triangle?

Ⓐ 45°
Ⓑ 60°
Ⓒ 90°
Ⓓ 180°

15. If ∠ A measures 45° and ∠ C measures 45° then what does ∠ B measure?

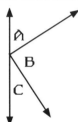

Ⓐ 45°
Ⓑ 75°
Ⓒ 90°
Ⓓ 135°

Measuring Turned Angles (4.MD.C.5.B)

1. At ice skating lessons, Erika attempts to do a 360 degree spin, but she only manages a half turn on her first attempt. How many degrees short of her goal was Erika's first attempt?

    Ⓐ 90 degrees
    Ⓑ 180 degrees
    Ⓒ 0 degrees
    Ⓓ 360 degrees

2. Erika's sister, Melanie, attempt to do a 360 degree turn, just like her sister, but she made a quarter turn on her first attempt. How many degrees short of her goal was Melanie's first attempt?

    Ⓐ 180 degrees
    Ⓑ 90 degrees
    Ⓒ 270 degrees
    Ⓓ 280 degrees

3. A water sprinkler covers 90 degrees of the Brown's backyard lawn. How many times will the sprinkler need to be moved in order to cover the full 360 degrees of the lawn?

    Ⓐ 4
    Ⓑ 2
    Ⓒ 3
    Ⓓ 5

4. A ceiling fan rotates 80 degrees and then it stops. How many more degrees does it need to rotate in order to make a full rotation?

    Ⓐ 265 degrees
    Ⓑ 90 degrees
    Ⓒ 180 degrees
    Ⓓ 280 degrees

Name _____   Date _____

5. Trixie is a professional photographer. She uses software on her computer to edit some wedding photos. She rotates the photograph of the bride 120 clockwise. Then she rotates it another 140 degrees clockwise. If she continues rotating the photo clockwise, how many more degrees will Trixie need to turn it to have made a complete 360 degree turn?

   Ⓐ 100 degrees
   Ⓑ 90 degrees
   Ⓒ 180 degrees
   Ⓓ 120 degrees

Name _____   Date _____

Measuring and Sketching Angles (4.MD.C.6)

1. What is the measure of angle JKL?

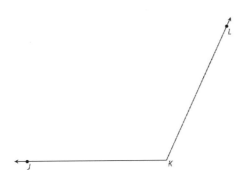

   Ⓐ 65 degrees
   Ⓑ 115 degrees
   Ⓒ 75 degrees
   Ⓓ 140 degrees

2. What is the measure of this angle?

   Ⓐ 10 degrees
   Ⓑ 28 degrees
   Ⓒ 48 degrees
   Ⓓ 90 degrees

Name _____   Date _____

3. What is the measure of angle PQR?

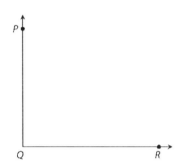

Ⓐ 360 degrees
Ⓑ 180 degrees
Ⓒ 0 degrees
Ⓓ 90 degrees

4. What is the measure of the interior angles of this shape?

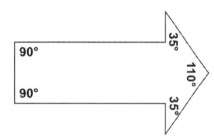

Ⓐ 400 degrees
Ⓑ 360 degrees
Ⓒ 540 degrees
Ⓓ 520 degrees

5. What can you use to help measure angles?

Ⓐ a degree
Ⓑ a protractor
Ⓒ a vertex
Ⓓ an angle

Adding and Subtracting Angle Measurements (4.MD.C.7)

1. Angle 1 measures 40 degrees, and angle 2 measures 30 degrees. What is the measure of angle of PQR?

   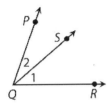

   - Ⓐ 70 degrees
   - Ⓑ 80 degrees
   - Ⓒ 100 degrees
   - Ⓓ 10 degrees

2. Angle ADC measures 120 degrees, and angle ADB measures 95 degrees. What is the measure of angle BDC?

   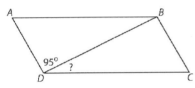

   - Ⓐ 5 degrees
   - Ⓑ 35 degrees
   - Ⓒ 15 degrees
   - Ⓓ 25 degrees

3. Angle JNM measures 100 degrees, angle JNK measures 25 degree, and angle KNL measures 35 degrees. What is the measure of angle LMN?

   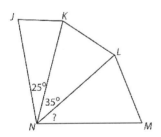

   - Ⓐ 35 degrees
   - Ⓑ 40 degrees
   - Ⓒ 30 degrees
   - Ⓓ 25 degrees

Name _____ Date _____

4. What fraction of a circle is a 180 degree angle?

   Ⓐ 1/4
   Ⓑ 1/2
   Ⓒ 1/3
   Ⓓ 1/5

5. What fraction of a circle is a 90 degree angle?

   Ⓐ 1/4
   Ⓑ 1/2
   Ⓒ 1/3
   Ⓓ 1/5

End of Measurement & Data

Name _____   Date _____

Measurement & Data

Answer Key
&
Detailed Explanations

Name _____   Date _____

Units of Measurement (4.MD.A.1)

Question No.	Answer	Detailed Explanation
1	A	Inches are not units of weight. Kilograms are not customary units. Tons are too large to use to measure a table. Pounds are the best choice.
2	B	Grams are not customary units. Between ounces and pounds, ounces are more reasonable unit to use.
3	A	12 lbs = 1.5 - 2 times a newborn baby. Gallons are used for liquid measurement. 200 ounces is only 12.5 pounds. 500 pounds is the most reasonable choice.
4	D	Meters and kilometers are not customary units. Yards are too small of a unit to measure the length of a road.
5	B	Gallons are not used to measure length. Centimeters are very small units. Miles are very large units. A tree's height would best be measured in yards.
6	A	Metric measurements are not customary units. Kilograms are part of the metric system.
7	B	Pounds measure weight, not liquid volume and yards measure length. Milliliters are part of the metric system, not the customary system of measurement.
8	C	There are 6 teaspoons in one fluid ounce. Multiply 6 x 2 to make a comparison.
9	B	There are 16 ounces in 1 pound. 125 oz ≈ 8 pounds (too heavy) 200 oz = 12.5 pounds (too heavy) 2 ounces is way too light for a textbook.
10	B	There are 3 teaspoons in 1 tablespoon.

Name _____ Date _____

Measurement Problems (4.MD.A.2)

Question No.	Answer	Detailed Explanation
1	B	To determine the difference between the time Arthur wants to arrive at practice and how long it takes him to get there, subtract 45 minutes from 5:30 by counting backwards by 5's on the clock:
2	A	Add 2 hours to 7:05, which would make 9:05. Count forward by 5's on the face of a clock for 35 minutes, then add 3 more minutes to determine the exact time:
3	B	There are 12 inches in 1 foot: Multiply 12 x 4, and then and add 5 more inches.
4	C	Add the prices of all the items and the amount of the taxes together. Subtract the sum from 300.00. $47.00 + 150.00 + 32.00 + 13.74 = $242.74 $300.00 - 242.74 = $57.26
5	C	One tablespoon equals 1/16 of a cup. Eight, 1/16 of a cup of equals 1/2 cup.
6	C	There are 16 ounces in 1 pound. Therefore, 32 ounces is more than 1 pound.

Name _____   Date _____

7	C	There are 12 months in 1 year. Calculate the total number of months and divide them by 12. The number of times 12 is divided by this number is the number of years, and the remainder is the number of months. 12 + 9 + 23 = 44 months 44 months = 3 years and 8 months
8	C	This is asking if the amount of flour used in the recipe is more or less than the amount of butter, and if the amount of butter is more or less than the amount of sugar. Check the given amounts to make a comparison.
9	B	The unit price is $3.25. Multiply that number by 4 to determine the cost of 4 books.
10	D	Multiply $24.99 x 3. Multiply $6.99 x 2. Add the products. $24.99 x 3 = $74.97 $6.99 x 2 = $13.98 $74.97 + 13.98 = $88.95

		Perimeter & Area (4.MD.B.3)
1	B	The formula for finding the area of a rectangle: Area = length x width.
2	A	The formula to find the perimeter of a rectangle: P = 2(length) + 2(width)
3	C	The length is 36 ft. and the the width is 48 ft.. All sides must be added to determine the perimeter. Add 48 + 36. Then multiply the sum by 2.
4	A	For the perimeter, add the measurements of the length and width: 72 + 30. Multiply the sum by 2.
5	A	Add the measurement of the length and the width, then multiply the sum by 2. Do this for both figures and compare the perimeters.
6	C	The perimeter is the sum of all 4 sides. If that sum is 100, divide 100 by 4 to determine the length of each side. (A square has 4 equal sides.)
7	D	To find the perimeter of a shape on the graph sheet, count each side of the squares all the way around the outside of the shape. The side of each square that makes up the shape is a unit.
8	B	To find the area of a shape on the graph sheet, count the number of squares on the inside of the shape.

Question No.	Answer	Detailed Explanation
9	C	Count the segments between the red dots, all the way around the shape.
10	B	Although all three of the choices offered would have an area of 48 square units, only a 12 by 4 rectangle would have a perimeter of 32 units.

Representing and Interpreting Data (4.MD.B.4)

Question No.	Answer	Detailed Explanation
1	D	The horizontal scale represents the number of cousins each student has. Read the number of xs plotted for 0 cousins.
2	C	Count the number of xs plotted above the number 4 on the horizontal scale.
3	B	The 2 foods that have the highest number of xs are the most favorites.
4	B	Subtract the number of families that pack for a picnic the night before from the number of families that pack before leaving for a picnic. Each x represents 2 points.
5	A	The horizontal scale represents the days of the week. The vertical scale represents the number of office visits. The line on the graph started at 20 and continued to increase to almost sixty.
6	B	The question is asking for a **difference** or comparison, which means **subtraction**. Find the number of visits for both days, then subtract: Monday = 20 from Friday = 55: 55 - 20.
7	B	Read the number that corresponds to the point plotted above Wednesday. If the point is halfway between numbers, add 5 points to the lower number the point sits between.
8	A	Add the total number of squares and triangles drawn on the graph. Each square represents 10 boxes and there are 14 squares shown: multiply 10 x 14. Each triangle represents 5 boxes and there are 3 triangles shown: 5 x 3. Calculate the products and add them together. Compare that sum to the total number of boxes needed to be sold.

Name _____   Date _____

Question No.	Answer	Detailed Explanation
9	B	To find the difference, total the number of boxes sold the first week, then the fourth week. Subtract the totals. Forty boxes were sold the first week; 55 boxes were sold the fourth week: 55 - 40.
10	D	To find the total, add the number of boxes sold both weeks. The first week, 40 boxes were sold; the second week, 35 boxes were sold: 40 + 35.
11	C	The horizontal scale represents the weeks that cans were sold. The vertical scale represents the number of cans sold. Compare the height of the bar to the number on the number scale for the second week. If a bar ends between two numbers, add 5 to the lower number the bar rests between.
12	B	Subtract to find the difference. Twenty-five cans were collected the second week; 40 were collected the third week: 40 - 25.
13	B	Compare the height of the bar to the number scale. Add 5 points to the lower number when the bar rests between numbers.
14	A	The week that has the shortest bar is the week with the fewest cans collected.
15	D	The weeks that show bars having the same height are the weeks that have the same number of collected cans.
16	C	The season showing the most tally marks is the most popular favorite season.
17	C	Count the number of tally marks shown for winter. Four tally marks with one tally crossing over them = 5 tally marks.
18	B	Subtract the number of tally marks shown for spring from the number of tally marks shown for winter.
19	B	To determine the total number of students who were surveyed or voted, add up the total number of tally marks shown on the graph.
20	C	Count the number of votes each field trip choice has. Add together the two field trip choices that have the most votes.
21	C	The range 10 - 50 for January is the greatest range shown on the chart.

Name _____   Date _____

Question No.	Answer	Detailed Explanation
22	B	The horizontal scale lists the years' studies about the number of people wearing bicycle helmets while bike riding have been done. The vertical scale shows the number of people who wear bicycle helmets while bike riding. One bar shows the number of injuries and deaths that occurred while not wearing a helmet. The other bar shows the number of bike riders wearing helmets. Compare the height of the bars to the number scale to determine the number of people wearing bike helmets in each time frame and the number of injuries or deaths. Then compare the difference between these two sets of data.
23	B	Find the sum of all the sweet fruit smoothie orders, then find the sum of all the sour fruit smoothie orders. Subtract the number of sour fruit smoothie orders from the sweet fruit smoothie orders to determine the difference.
24	D	6-5 1/4= 3/4
25	D	Convert 60 minutes into 1 hour. There are 3 students that spent 1 hour, but only 1 that spend more time than that.
26	D	5 2/3-1= 4 2/3
27	B	"Jump" the line from "0" to "1". The denominator will be the number of jumps it takes to get to "1". The numerator will be how many "jumps" to the end of the pencil.
28	C	"Jump" the line from "0" to "1". The denominator will be the number of jumps it takes to get to "1". The numerator will be how many "jumps" to the end of the pencil.
29	D	"Jump" the line from "0" to "1". The denominator will be the number of jumps it takes to get to "1". The numerator will be how many "jumps" to the end of the pencil.
30	B	Find 3/4 on the plot, and count the number of x's above that increment.

Angle Measurement (4.MD.C.5.A)

Question No.	Answer	Detailed Explanation
1	B	Angle C and Angle A are congruent. They have the same exact measurements.
2	A	Angle C and Angle D are supplementary. Their measures total 180°. 180 - 128 = 52
3	D	Angle B and Angle D are congruent so they will measure the same.
4	C	Together, Angle A and Angle B equal 180 degrees. Therefore, subtract 64 degrees from 180 to determine the measure of Angle B.
5	B	Subtract 94 from 180 to find the measure of Angle A.
6	D	To find the measure of Angle B, subtract 51 from 90.
7	C	Two supplementary angles together form a straight angle, which is equal to half of a circle and measures 180 degrees.
8	B	Two complementary angles total 90 degrees.
9	C	A right angle measures 90 degrees and looks like this: This is an example of an obtuse angle: Obtuse Angle 4.MD.5.a
10	C	Subtract 7 from 90. 90 - 7 = 83
11	A	Angle A and Angle B are supplementary. Subtract 46 from 180.
12	D	The measures of these three angles total 180 degrees. Add the measures of Angles A and B. Subtract the sum of those two angles from 180 to determine how much Angle C measures. 101 + 49 = 150 180 - 150 = 30
13	B	Subtract 73 from 180. 180 - 73 = 107

Name _____ Date _____

Question No.	Answer	Detailed Explanation
14	B	This is an equilateral triangle, in which all interior angles have the exact same measurement. Those angles total 180 degrees. 180 ÷ 3 = 60. Each angle measures 60 degrees.
15	C	All the angle measures must add up to 180 degrees. Add 45 and 45. Subtract the sum from 180. (Angle B is a right angle.)

Measuring Turned Angles (4.MD.C.5.B)

1	B	A half-turn would be 180 degrees. Erika made a 180 degree turn, but her goal was 360 degrees. 360-180= 180 degrees.
2	C	Melanie's turn was 90 degrees. 360 (her goal) minus 90 (her turn) equals 270 degrees.
3	A	360 divided by 90 is 4. The sprinkler will need to be moved 4 times in order to cover the lawn.
4	D	360 degrees minus 80 degrees equals 280 degrees.
5	A	120 degrees plus 140 degrees equals 260 degrees. 360 degrees minus 260 degrees leaves 100 degrees left to make a full turn.

Measuring and Sketching Angles (4.MD.C.6)

1	B	Put the protractor with its center point on K, so that one ray points to 0°, the other ray points to 115°.
2	B	Put the protractor with its center on the vertex and one ray pointing to 0°, so the other ray points to 28°.
3	D	Put the protractor with its center on the vertex and one ray pointing to 0°, so the other ray points to 90°.
4	C	Each quadrilateral has 360 degrees, and all triangles have 180 degrees, so you can divide the arrow polygon into two non-overlapping shapes--a rectangle and a triangle. The angle measurement inside the arrow shape was 540 degrees total (add 360 degrees + 180 degrees).
5	B	Definition of protractor: a tool used to measure angles.

Adding and Subtracting Angle Measurements (4.MD.C.7)

Question No.	Answer	Detailed Explanation
1	A	The measure of the angle PQR is the sum of angle 1 & angle 2, so it is 40 + 30 = 70.
2	D	The measure of angle BDC can be found by subtracting 95 degrees from 120 degrees, for 25 degrees
3	B	Subtract 35 degrees from 100 degrees, which equals 40 degrees.
4	B	1/2 of a circle is 180 degrees.
5	A	1/4 of a circle is 90 degrees.

Name _____   Date _____

Geometry

Section 1: Draw and identify lines and angles, and classify shapes by properties of their lines and angles.

Points, Lines, Rays, and Segments (4.G.A.1)

1. Which of the following is a quadrilateral?

   Ⓐ Triangle
   Ⓑ Rhombus
   Ⓒ Pentagon
   Ⓓ Hexagon

2. How many sides does a pentagon have?

   Ⓐ 3
   Ⓑ 2
   Ⓒ 1
   Ⓓ 5

3. Use the network below to respond to the following question:
   How many line segments connect directly to Vertex F?

   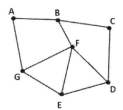

   Ⓐ 3
   Ⓑ 4
   Ⓒ 5
   Ⓓ 6

4. Which of these statements is true?

   Ⓐ A parallelogram must be a rectangle.
   Ⓑ A trapezoid might be a square.
   Ⓒ A rhombus must be a trapezoid.
   Ⓓ A rectangle must be a parallelogram.

5. What is being shown below?

   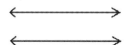

   Ⓐ a pair of parallel lines
   Ⓑ a pair of intersecting lines
   Ⓒ a pair of congruent rays
   Ⓓ a pair of perpendicular lines

6. What is being shown below?

   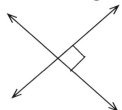

   Ⓐ a pair of parallel lines
   Ⓑ a pair of perpendicular lines
   Ⓒ a pair of obtuse angles
   Ⓓ a net for a cube

7. Identify the following plane figures.

   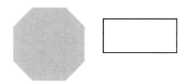

   Ⓐ heptagon and quadrilateral
   Ⓑ hexagon and quadrilateral
   Ⓒ octagon and quadrilateral
   Ⓓ octagon and pentagon

Name _____   Date _____

8. How many sides does an octagon have?

   Ⓐ 8
   Ⓑ 7
   Ⓒ 6
   Ⓓ 5

9. What kind of lines intersect to make 90-degree angles?

   Ⓐ supplementary
   Ⓑ perpendicular
   Ⓒ skew
   Ⓓ parallel

10. How many right angles does a parallelogram have?

    Ⓐ 0
    Ⓑ 2
    Ⓒ 4
    Ⓓ It depends on the type of parallelogram.

Name _____   Date _____

Angles (4.G.A.1)

1. The hands of this clock form a(n) _____ angle.

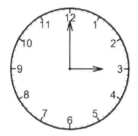

   Ⓐ obtuse
   Ⓑ straight
   Ⓒ right
   Ⓓ acute

2. How many acute angles and obtuse angles are there in the figure shown below?

   Ⓐ 2 acute angles and 6 obtuse angles
   Ⓑ 4 acute angles and 4 obtuse angles
   Ⓒ 8 acute angles and 0 obtuse angles
   Ⓓ 0 acute angles and 8 obtuse angles

3. Describe the angles found in this figure.

   Ⓐ 2 right angles and 3 acute angles
   Ⓑ 3 right angles and 2 obtuse angles
   Ⓒ 2 right angles, 2 obtuse angles, and 1 acute angle
   Ⓓ 2 right angles and 3 obtuse angles

4. What type of triangle is shown below?

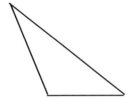

   Ⓐ Isosceles triangle
   Ⓑ Scalene triangle
   Ⓒ Equilateral triangle
   Ⓓ None of the above

5. Which statement is true about an obtuse angle?

   Ⓐ It measures less than 90 degrees.
   Ⓑ It measures more than 90 degrees.
   Ⓒ It measures exactly 90 degrees.
   Ⓓ It measures more than 180 degrees.

6. Which statement is true about a straight angle?

   Ⓐ It measures less than 90 degrees.
   Ⓑ It measures more than 90 degrees.
   Ⓒ It measures exactly 90 degrees.
   Ⓓ It measures exactly 180 degrees.

7. A square has what type of angles?

   Ⓐ 2 acute and 2 right angles
   Ⓑ 4 right angles
   Ⓒ 4 acute angles
   Ⓓ It depends on the size of the square.

8. Which statement is true about an equilateral triangle.

   Ⓐ It has 1 acute angle and 2 obtuse angles.
   Ⓑ It has 2 acute angles and 1 right angle.
   Ⓒ It has all 30-degree angles.
   Ⓓ It has all 60-degree angles.

9. Classify the angle:

   Ⓐ right
   Ⓑ acute
   Ⓒ obtuse
   Ⓓ straight

10. Tina wanted her bedroom area rug to be designed after a geometric shape. The rug has somewhat of a circular shape with 7 straight sides and 7 obtuse angles. What is the name of this shape?

    Ⓐ hexagon
    Ⓑ octagon
    Ⓒ pentagon
    Ⓓ heptagon

Name _____ Date _____

Classifying Plane (2-D) Shapes (4.G.A.2)

1. Complete the sentence:
   A polygon is named based on _____.

   Ⓐ how many sides or interior angles it has
   Ⓑ how many of its sides are straight
   Ⓒ how large it is
   Ⓓ how many lines of symmetry it has

2. Complete the sentence:
   A polygon with 4 sides and 4 interior angles is called a(n) _____.

   Ⓐ triangle
   Ⓑ quadrilateral
   Ⓒ pentagon
   Ⓓ octagon

3. Complete the sentence:
   A rectangle must have _____ .

   Ⓐ all parallel sides and all congruent sides
   Ⓑ 2 pairs of parallel sides and 2 pairs of congruent sides
   Ⓒ 2 pairs of parallel sides and 4 congruent sides
   Ⓓ 4 parallel sides and 2 pairs of congruent sides

4. Complete the sentence:
   A polygon must have _____ .

   Ⓐ 3 or more interior angles
   Ⓑ at least one pair of parallel sides
   Ⓒ at least one pair of congruent sides
   Ⓓ a line of symmetry

Name _____ Date _____

5. Classify this triangle:

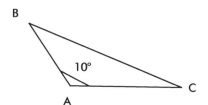

Ⓐ acute triangle
Ⓑ obtuse triangle
Ⓒ right triangle
Ⓓ straight triangle

6. If the angles on a triangle all measure less than 90 degrees, what type of triangle is it?

Ⓐ obtuse triangle
Ⓑ right triangle
Ⓒ straight triangle
Ⓓ acute triangle

7. Another name form a 180-degree angle is a(n) _____.

Ⓐ right angle
Ⓑ obtuse angle
Ⓒ straight angle
Ⓓ acute angle

8. Complete the sentence:
A point has _____.

Ⓐ size but no position
Ⓑ size and position
Ⓒ neither size nor position
Ⓓ position but no size

9. Which type of angle shows up the most in this photo?

   Ⓐ This photo mainly consists of acute angles.
   Ⓑ This photo mainly consists of straight angles.
   Ⓒ This photo mainly consists of right angles.
   Ⓓ This photo mainly consists of obtuse angles.

10. Classify the triangles shown in this design.

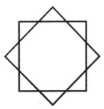

   Ⓐ They are acute triangles.
   Ⓑ They are right triangles.
   Ⓒ They are obtuse triangles.
   Ⓓ They are straight triangles.

Symmetry (4.G.A.3)

1. How many lines of symmetry does an equilateral triangle have?

    Ⓐ 1
    Ⓑ 3
    Ⓒ 2
    Ⓓ 0

2. How many lines of symmetry does a regular pentagon have?

    Ⓐ 1
    Ⓑ 2
    Ⓒ 5
    Ⓓ 10

3. How many lines of symmetry does a rectangle have? (Assume the rectangle is not also a square.)

    Ⓐ 1
    Ⓑ 2
    Ⓒ 3
    Ⓓ 4

4. How many lines of symmetry does a regular octagon have?

    Ⓐ 2
    Ⓑ 4
    Ⓒ 6
    Ⓓ 8

5. How many lines of symmetry does a square have?

    Ⓐ 2
    Ⓑ 4
    Ⓒ 6
    Ⓓ 8

Name _____   Date _____

6. How many lines of symmetry does a parallelogram have if it is not a square, rectangle, or rhombus?

Ⓐ 2
Ⓑ 4
Ⓒ 1
Ⓓ 0

7. How many lines of symmetry does the following object have?

Ⓐ 0
Ⓑ 2
Ⓒ 4
Ⓓ 6

8. How many lines of symmetry does the following object have?

Ⓐ 0
Ⓑ 1
Ⓒ 2
Ⓓ 3

9. How many lines of symmetry does the green portion of the following shape have?

Ⓐ 4
Ⓑ 3
Ⓒ 2
Ⓓ 1

10. Regular rectangles have 4 angles and 4 lines of symmetry. Regular pentagons have 5 angles and 5 lines of symmetry. Regular hexagons have 6 angles and 6 lines of symmetry. If this pattern were to continue, how many lines of symmetry does a regular heptagon have?

   Ⓐ 7
   Ⓑ 8
   Ⓒ 3
   Ⓓ 2

End of Geometry

Geometry

Answer Key
&
Detailed Explanations

Points, Lines, Rays, and Segments (4.G.A.1)

Question No.	Answer	Detailed Explanation
1	B	Quadrilaterals are 4 sided-polygons. The prefix "quad" means 4. A rhombus has 4 sides.
2	D	The prefix "pent" means 5.
3	B	Vertices are points where two line segments meet, or intersect. Four line segments meet at Point F.
4	D	Vertices are points where two line segments meet, or intersect. Four line segments meet at Point F.
5	A	Lines are drawn with arrows on each end. If they are perpendicular or intersecting, they cross each other. These two lines will never cross. They are called parallel.
6	B	These two lines cross to form right angles. By definition, the lines are perpendicular.
7	C	A polygon is named after the number of sides and interior angles it has: "tri" means 3, "quad" means 4, "pent" means 5, "hex" means 6, "hept" means 7 and "oct" means 8.
8	A	The prefix in front of the word octagon is "oct" and means 8.
9	B	By definition, perpendicular lines must cross to form right (90-degree) angles.
10	D	A parallelogram is a quadrilateral with two pairs of parallel sides. There is no requirement for it to have a certain number of right angles. It may have 0 or it may have 4.

Name _____   Date _____

Angles (4.G.A.1)

Question No.	Answer	Detailed Explanation
1	C	The hands of this clock differ by 90 degrees.
2	D	Obtuse angles are larger than 90 degrees; acute angles are smaller than 90 degrees.
3	D	Right angles are 90 degrees, acute angles are less than 90 degrees and obtuse angles are more than 90 degrees.
4	B	This triangle has no equal sides. An equilateral triangle has all equal sides; an isosceles triangle has 2 equal sides.
5	B	An obtuse angle measures between 90 and 180 degrees.
6	D	A straight angle measures the same as 2 right angles. It is half of a full circle, which is 360 degrees.
7	B	Each angle of a square measure 90 degrees, regardless of its size.
8	D	An equilateral triangle is a closed shape that has all equal angles that add up to 180 degrees. This means each of the three angles measures 60°.
9	B	This angle measures less than 90 degrees.
10	D	The prefix "hept" means 7. A heptagon is a polygon with 7 sides and 7 angles.

Classifying Plane (2-D) Shapes (4.G.A.2)

1	A	The names given to different types of polygons (triangle, quadrilateral, pentagon, etc.) are based on how many sides or interior angles each figure has.
2	B	The prefix "quad' means 4.
3	B	The opposite sides of a rectangle must be congruent and parallel.
4	A	Triangles, quadrilaterals, pentagons, hexagons, heptagons and octagons are examples of polygons. They all have at least 3 interior angles.
5	B	This triangle has an angle that measures more than 90 degrees. Therefore, it is classified as an obtuse triangle.

Question No.	Answer	Detailed Explanation
6	D	An acute triangle has three angles that each measure less than 90 degrees.
7	C	A straight angle is an angle that measures exactly 180 degrees.
8	D	A point is just a location. It has a position in space, but no size.
9	C	Many of the angles in this photo measure 90 degrees. There are mostly right angles (the corners of the house, the door and window frames, the steps in the pool).
10	B	The triangles in this design have right angles. Therefore, they would be classified as right triangles.

Symmetry (4.G.A.3)

Question No.	Answer	Detailed Explanation
1	B	A line of symmetry is an imaginary line that separates a figure into identical parts.
2	C	For **regular** polygons, the number of lines of symmetry equals the number of sides the shape has. A pentagon has 5 sides, so there are 5 lines of symmetry.
3	B	A rectangle has two lines of symmetry. One is horizontal through its center, the other is vertical through its center.
4	D	An octagon is an 8-sided polygon. Therefore, a regular octagon would have 8 lines of symmetry.
5	B	A square is actually a regular quadrilateral. Therefore, it must have four lines of symmetry.
6	D	Irregular parallelograms are slanted. Therefore, the two sides will not match up when folded. There are no lines of symmetry.
7	C	A line of symmetry divides a figure into identical parts or mirror images of each other. This figure has one horizontal, one vertical, and two diagonal lines of symmetry.
8	B	One line can be drawn vertically through the center of this shape to divide it into two matching parts.
9	D	Both **vertical** sides of this figure have the same shape. The point at the top is perfectly centered.
10	A	Each mentioned polygon increases by one angle and line of symmetry. The prefix "hept" means 7.

Notes

Lumos StepUp™ is an educational app that helps students learn and master grade-level skills in Math and English Language Arts.

The list of features includes:

- Learn Anywhere, Anytime!
- Grades 3-8 Mathematics and English Language Arts
- Get instant access to the Common Core State Standards
- One full-length sample practice test in all Grades and Subjects
- Full-length Practice Tests, Partial Tests and Standards-based Tests
- 2 Test Modes: Normal mode and Learning mode
- Learning Mode gives the user a step-by-step explanation if the answer is wrong
- Access to Online Workbooks
- Provides ability to directly scan QR Codes
- And it's completely FREE!

http://lumoslearning.com/a/stepup-app

About Online Workbooks

- When you buy this book, 1 year access to online workbooks is included

- Access them anytime from a computer with an internet connection

- Adheres to the Common Core State Standards

- Includes progress reports

- Instant feedback and self-paced

- Ability to review incorrect answers

- Parents and Teachers can assist in student's learning by reviewing their areas of difficulty

Course Name: Grade 4 Math Prep

Lesson Name:	Correct	Total	% Score	Incorrect
Introduction				
Diagnostic Test		3	0%	3
Number and Numerical Operations				
Workbook - Number Sense	2	10	20%	8
Workbook - Numerical Operations	2	25	8%	23
Workbook - Estimation	1	3	33%	2
Geometry and measurement				
Workbook - Geometric Properties		6	0%	6
Workbook - Transforming Shapes				
Workbook - Coordinate Geometry	1	3	33%	2
Workbook - Units of Measurement				
Workbook - Measuring Geometric Objects	3	10	30%	7
Patterns and algebra				
Workbook - Patterns	7	10	70%	3
Workbook - Functions and relationships				

LESSON NAME: Workbook - Geometric Properties
Elapsed Time: 01:19

Question No. 2
What type of motion is being modeled here?

Select right answer
- ○ a translation
- ○ a rotation 90° clockwise
- ● a rotation 90° counter-clockwise
- ○ a reflection

[Previous question] [Next question]

Report Name: Missed Questions
Student Name: Lisa Colbright
Cours Name: Grade 4 Math Prep
Lesson Name: Diagnostic Test

The faces on a number cube are labeled with the numbers 1 through 6. What is the probability of rolling a number greater than 4?

Answer Explanation

(C) On a standard number cube, there are six possible outcomes. Of those outcomes, 2 of them are greater than 4. Thus, the probability of rolling a number greater than 4 is "2 out of 6" or 2/6.

A)   1/6
B)   1/3
C) Correct Answer   2/6
D)   3/6

Lumos Learning
Developed By Expert Teachers

Common Core Practice
ENGLISH LANGUAGE ARTS

Grade 4

(((tedBook)))

★ Three Strands

★ Hundreds of Activities

30+ SKILLS

PLUS **Online Workbooks**

Foundational Skills for
PARCC or
Smarter Balanced Tests

Available
- At Leading book stores
- Online www.LumosLearning.com

Made in the USA
San Bernardino, CA
06 December 2015